Personal Web Usage in the Workplace: A Guide to Effective Human Resources Management

Murugan Anandarajan
Drexel University, USA

Claire A. Simmers
Saint Joseph's University, USA

Information Science Publishing
Hershey • London • Melbourne • Singapore

Acquisition Editor: Mehdi Khosrow-Pour
Senior Managing Editor: Jan Travers
Managing Editor: Amanda Appicello
Development Editor: Michele Rossi
Copy Editor: Maria Boyer
Typesetter: Jennifer Wetzel
Cover Design: Michelle Waters
Printed at: Integrated Book Technology

Published in the United States of America by
Information Science Publishing (an imprint of Idea Group Inc.)
701 E. Chocolate Avenue, Suite 200
Hershey PA 17033
Tel: 717-533-8845
Fax: 717-533-8661
E-mail: cust@idea-group.com
Web site: http://www.idea-group.com

and in the United Kingdom by
Information Science Publishing (an imprint of Idea Group Inc.)
3 Henrietta Street
Covent Garden
London WC2E 8LU
Tel: 44 20 7240 0856
Fax: 44 20 7379 3313
Web site: http://www.eurospan.co.uk

Library of Congress Cataloging-in-Publication Data

Personal web usage in the workplace : a guide to effective human resources management / Murugan Anandarajan, Claire A. Simmers, editors.
p. cm.
ISBN 1-59140-148-8
1. Personal Internet use in the workplace. I. Anandarajan, Murugan, 1961- II. Simmers, Claire, 1950-
HF5549.5.P39P47 2003
658.3'12--dc22

2003014951

eISBN 1-59140-149-6
paperback ISBN 1-59140-287-5

British Cataloguing in Publication Data
A Cataloguing in Publication record for this book is available from the British Library.

All work contributed to this book is new, previously-unpublished material. The views expressed in this book are those of the authors, but not necessarily of the publisher.

Dedications

To my beloved parents and aunt, your belief in me is truly inspirational - MA

To Michael, Jessica, and Christa, always there with love and support - CAS

Personal Web Usage in the Workplace: A Guide to Effective Human Resources Management

Table of Contents

Section III: Toward the Well-Being of the Employee

Preface

Few will deny that the increasingly omnipresent nature of the World Wide Web in the workplace is dramatically revolutionizing the manner in which we work. The advantages of the World Wide Web are the ability to gather, communicate, distribute, share, and store information publicly in real time (Davis & Naumann, 1999). The reach and range of the World Wide Web is phenomenal (Evans & Wurster, 2000) and employees have increasingly been given access to it in the workplace.

Employees also view the World Wide Web as an indispensable tool, using it to communicate with colleagues, managers, and subordinates, and to maintain relationships with valued customers. According to the UCLA Internet Report, Surveying the Digital Future, Year 3 (2003, p. 72), of those who had Internet access at work, 90% visited work-related sites in 2002, up from 89% in 2001 and 83% in 2000. There is some evidence that the Internet is perceived as a catalyst for productivity, while those who report that the Internet makes them neither more nor less productive continue to decline (UCLA Center for Communication Policy, 2003, p. 75).

In addition to being an organizational tool, the Web provides employees access to the world's biggest playground and information repository. This aspect has prompted growing concerns about personal World Wide Web usage in the workplace. According to IDC Research, 30% to 40% of employee World Wide Web activity is non-business-related. The UCLA Internet Report, Surveying the Digital Future, Year 3 reports that of those who had Internet access at work, about 60% visited websites for personal use in 2002, about the same as in 2001.

Since the World Wide Web is an integral component of our workplaces, then management of personal use is a timely topic. There seems to be two major perspectives framing the management of personal Web usage (PWU) in the workplace. The first is that PWU is dysfunctional. It is negative, with no place in the workplace, as it can cost organizations billions of dollars in terms of lost productivity, increased security costs, and network overload, as well as the risk of civil and criminal liabilities. Personal usage at work is depicted as a variation of other dysfunctional work behaviors such as stealing, wasting time, and making personal long distance phone calls (Block, 2001). In this perspective PWU is often called cyber slacking, or Web abuse, or cyber deviance. This perspective fosters the characterization of employees as "variable costs" that are to be monitored, controlled, and where possible, minimized; it is more of an adversarial view of the employment relationship. To monitor and control personal Web usage, organizations often use information technology control mechanisms such as firewalls, content management software, log files, and blocking (Sunoo, 1996).

A second viewpoint is that PWU has the potential for constructive effects; roots of this viewpoint are in a human resource perspective. A human resource perspective views employees as valuable assets that are to be nurtured and invested in. This perspective considers employees as partners where collaboration and trust are the drivers of organizational and personal interfaces. When employees are viewed as investments, there are incentives to invest in such things as training, development, prevention of skill obsolescence, retention programs, wellness, and work life balance because the returns to these investments, less immediate and tangible, are real. The human resource perspective is of increasing importance in the 21st century workplace because it is provides a stronger foundation for competitive advantage than products and facilities, which are easily imitated. A human resource-based view of the firm suggests that sustainable advantage derives primarily from human skills, knowledge bases, and service strengths that are not easily reproduced (Quinn, Doorley, & Paquette, 1990), and there is recognition that having superior people in your organization is critical. Personal Web usage then can have learning and well-being components from a human resource view.

Personal Web usage can contribute to the continuous learning so important for 21st century "knowledge workers." The Web can be used to keep current on world events and business news, and to support educational efforts through formal classes and professional associations. As examples of the well-being component, PWU can be a way to manage an increasingly blended work and personal life. PWU permits the accomplishment of personal tasks that have been displaced as work demands spread out beyond the traditional eight-hour day, five-day-a-week work schedule. Surprisingly, in a recent survey it was discovered that Americans spend more time at home on the Internet for work purposes than they spend on the Internet at work for personal reasons (Kaplan, 2003). Allowing PWU in the workplace then would seem to be equitable repayment for work done at home. Additionally, PWU might foster subconscious problem solving or provide a necessary break from drudgery or intense endeavor..." (Friedman, 2000, p. 1563).

The paradox then is how to blend the control perspective with reliance on hard controls through impersonal information technologies with the human resource perspective with reliance on interpersonal communication, and a shared understanding of acceptable Internet behaviors. This volume presents work that focuses on understanding and resolving this paradox.

ORGANIZATION OF THIS BOOK

Information Systems has become a wide and diverse discipline as information technology has moved from back-office, closed systems to end-user-controlled open systems. To fully appreciate the role of information technology in the 21st century workplace requires a range of approaches. However, in this volume, we have chosen to explore one aspect of information technology — personal Web use in the workplace through the lens of the human resource view. We feel that successful organizations in the 21st century will be those that attract, retain, develop, and reward individuals who have skills and knowledge to creatively approach customers, stakeholders, and take advantage of the opportunities that the World Wide Web offers in a global marketplace.

In the first section, *"Exploring the Paradox of Personal Web Usage,"* the positive and negative aspects of PWU are examined. In Chapter 1, Murugan Anandarajan and Claire Simmers present the results of a qualitative study in which two dimensions of personal Web usage (constructive and dysfunctional) are identified. They find that organizational position is an important factor influencing judgments on the appropriateness of PWU. Chapter 2, by Zoonky Lee, Younghwa Lee, and Yongbeom Kim, examines why employees use the Internet for personal purposes during work hours. Employees use the Web for personal use because they do not think it is harmful or unethical, because of strong social influence, and because PWU may be beneficial to the organization. The main deterrents to PWU are lack of time and lack of privacy. Jo Ann Oravec in Chapter 3 proposes that constructive uses of online recreation and play can enhance many workplaces (especially high-tech and information-saturated ones), helping individuals gain fresh perspectives. She suggests that workgroups and human resource professionals participate in discussions as to what constitutes "constructive recreation" and in the development of fair organizational policies. In the last chapter of this section, Murugan Anandarajan, Patrick Devine, and Claire Simmers use multidimensional scaling techniques to develop a typology of workplace personal Web usage, with PWU behaviors falling into four distinct categories: disruptive, recreational, personal learning, and ambiguous.

In the chapters in the second section, *"Managing Personal Web Usage from a Human Resource Perspective,"* the range of options available to manage PWU is explored. Susan Lippert addresses the concept and importance of interpersonal trust and the use of the Internet in an organizational setting. Generalized guidelines for organizational practice and recommendations to support a culture of trust within the work environment are presented. In Chapter 6, Dinesh Mirchandani draws from the field of criminology using deterrence theory to investigate PWU. Deterrence theory suggests that sanctions and disincentive measures can reduce systems abuse by making potential abusers aware that their unethical behavior will be detrimental to their own good. Mirchandani recommends that a human resource manager, rather than an information technology person, spearhead organizational efforts handling PWU in the organization.

Chapter 7 by Paulette Alexander takes a different view by looking at how employees are subjected to unsolicited Web intrusions that may be interpreted as dysfunctional PWU. Alexander recommends policies and practices in addition to the deployment of protective technologies to shield both employees and the organization. Andrew Urbaczewski in Chapter 8 provides a classification and description of various control mechanisms, both technical and social. The social solutions rely on interpersonal skills rather than the "hammer of the log file" to curb dysfunctional personal Web usage. In Chapter 9, Claire Simmers and Murugan Anandarajan examine whether employee web usage patterns, attitudes toward web usage in the workplace, and organizational policies are more similar (convergence thesis) or less similar (divergence thesis) in three countries. The section concludes with Chapter 10, where Grania Connors and Michael Aikenhead examine the legal implications of PWU in the workplace for both employees and employers. In the United States, the significant risks to which employers are exposed outweigh an individual's right to privacy.

The final section is entitled *"Toward the Well-Being of the Employee."* In Chapter 11, Feng-Yang Kuo discusses Internet abuse from a psychoanalytic perspective. While past research has treated abuse as deriving from conscious decision, the unconscious mind may influence one's abusive conduct.

Thus social responsibilities and sanctions, and individual psychological well-being should be part of the training process in organizations as much as technical training. In Chapter 12, Mark Griffiths continues to examine the issue of employee well-being from a different lens by introducing the concept of Internet addiction, specifically looking at online pornography, sexually related Internet crime, and online gambling in the workplace. He offers guidelines for employers and human resource departments such as raising awareness, partnering with employees so everyone is vigilant, and giving support and help to problem users. The final chapter is written by Pruthikrai Mahatanankoon and Magid Igbaria who found that personal e-commerce enhanced job satisfaction and productivity, while personal information seeking decreased productivity. They suggest that attitudinal changes and enforced behavioral norms developed through education and training, rather than relying on filtering, and monitoring tools show the most promise for managing personal Web usage in the workplace.

This book continues to add to our body of knowledge on personal Web usage in the workplace and supports viewing the issue from a human resource perspective. As organizations look to employees as the competitive key, then how PWU is managed is one indicator of how seriously an organization takes the mission of the human resource perspective to heart and to practice.

REFERENCES

Block, W. (2001). Cyberslacking, business ethics and managerial economics. *Journal of Business Ethics,* 33(3), 225-231.

Evans & Wurster (2000). *Blown to Bits*. Boston, MA: Harvard Business School Press.

Friedman, W.H. (2000). Is the answer to Internet addiction, Internet interdiction? In Chung, M. (Ed.), *Proceedings of the 2000 Americas Conference on Information Systems.*

Kaplan, D. (2003). Work habits. *Adweek Eastern Edition*, 44(8), 37.

Quinn, J.B., Doorley, T.L., & Paquette, P.C. (1990). Beyond products: Service-based strategy. *Harvard Business Review*, 90(2), 58-67.

Sunoo, B.P. (1996). The employee may be loafing. *Personnel Journal*, (December), 55-62.

UCLA Center for Communication Policy. (2003). *The UCLA Internet Report — Surveying the Digital Future*. Accessed March 28, 2003, from: http://www.ccp.ulca.edu.

Acknowledgments

Books of this nature are written only with the support of many individuals. We would like to thank the book's contributors, all of whom generously shared their vast knowledge of Web usage with us. We would like to acknowledge the help of all involved in the review process of the book, without whose support the project could not have been satisfactorily completed. A further special note of thanks goes also to the publishing team at Idea Group Publishing. In particular to Michele Rossi and Jennifer Sundstrom, both who continuously kept in touch, keeping the project on schedule, as well as to Mehdi Khosrow-Pour, whose enthusiasm motivated us to initially accept his invitation for taking on this project. In addition, we would like to thank Drexel University graduate students, Shilpa Ramdas Mahangade, Gaurav Wason, and Maliha Zaman who helped in administrating the entire process.

Finally, we thank our families, Sharmini, Vinesh, Dharman and Michael, Jessica, and Christa, for their love and support throughout this project.

Murugan Anandarajan, PhD
Department of Management
Drexel University, USA

Claire A. Simmers, PhD
Department of Management
Saint Joseph's University, USA

Section I

Exploring the Paradox of Personal Web Usage

Chapter I

Constructive and Dysfunctional Personal Web Usage in the Workplace: Mapping Employee Attitudes

Murugan Anandarajan
Drexel University, USA

Claire A. Simmers
Saint Joseph's University, USA

ABSTRACT

In order to better understand how people work in the Web-enabled workplace, we examined the phenomenon of personal Web usage (PWU). We analyzed 316 responses from those with Web access at work to the question, "Do you think it's ok for a person to use the Web for non-work purposes during working hours in the workplace." The responses were coded into 19 themes and four categories. Using correspondence analysis, concept maps were generated which revealed that personal Web usage in the workplace is a complex issue with not only a potentially dysfunctional dimension, but also a potentially constructive one. Organizational position was an important variable with top, middle, lower-level managers, as well as professionals, and administrators positioning in different spaces on the conceptual map. Further analysis using Q-methodology reinforced the dual nature of PWU and the importance of position. Drawing on our results, an extension of the social contract theory and a model of personal Web usage in the workplace were suggested.

INTRODUCTION

"The Internet has brought distractions into cubicles ... Employee study cites rampant Internet abuse." (Network World, 2000)

Such headlines have become familiar popular press items. According to the American Management Association, more than 50% of all workplace-related Web activities are personal in nature (Greengard, 2000). Examples of personal Web usage (PWU) activities include reading news, making travel arrangements, online purchases, and searching for jobs. Personal Web usage has consistently been seen as a negative force with productivity losses, congested computer resources, security costs, and legal liability risks prominent concerns (Conlin, 2000). As the business environment becomes increasingly Web-enabled, organizations show a growing interest in understanding and managing PWU (McWilliams & Stepanek, 1998; Stewart, 2000; Simmers, 2002).

Personal Web usage has been defined as any voluntary act of employees using their company's Web access during office hours to surf non-work-related websites for non-work purposes (Lim et al., 2002). There seems to be three views on the issue of PWU. It is often assessed as completely negative, with no place in the workplace as it can cost organizations billions of dollars in terms of lost productivity, increased security costs and network overload, as well as the risk of civil and criminal liability. Another view is that personal usage at work is a variation of dysfunctional work behaviors such as stealing, wasting time, and making personal long distance phone calls. These behaviors need to be managed and controlled, primarily through monitoring, policies, and disciplinary actions (Block, 2001; Sunoo, 1996). In these two views, PWU is often called *cyber slacking* or *Web abuse.* However, a third view is that such "cyber activity, which might foster subconscious problem solving or provide a necessary break from drudgery or intense endeavor…might increase productivity" (Friedman, 2000, p. 1563). PWU might be viewed in the same light as an 'office-toy' such as clay, putty, or foam balls which are shown to decrease work stress and inspire creativity (Terr, 1999). Additionally, PWU can be a way to manage an increasingly blended work and personal life. PWU permits the accomplishment of personal tasks that have been displaced as work demands spread out beyond the traditional eight-hour day, five-day-a-week work schedule. Finally, PWU could contribute to the continuous learning that all employees are being called to as 21st century "knowledge workers."

The widespread prevalence of PWU and the general lack of understanding about it necessitate a systematic examination of the phenomenon. To date, relatively few empirical studies have addressed the issue of PWU in the workplace. The information systems literature has shown disproportionate emphasis behaviors such as the corporate benefits of Web usage (Anandarajan et al., 2000; Lederer et al., 2000; Teo & Lim, 1998) and, on the dark side of Web usage behavior (Griffiths, 1998; Joinson, 1998; Putnam & Maheu, 2000), identifying the types of websites accessed (Anandarajan et al., 2000; Teo et al., 1999) and on the time spent on such activity (Armstrong et al., 2000; Korgaonkar & Wolin, 1999; Teo et al., 1999). We have to yet to understand the underlying attitudes that influence such personal Web usage behaviors. This focus is consistent with the theory of reasoned action, which posits that attitudes can influence subsequent behavior both indirectly through influencing intention (Fishbein & Ajzen, 1975) and directly (Bentler & Speckart, 1981).

Specifically, the purpose of this study was threefold: (i) to explore employees' attitudes on PWU, (ii) to identify underlying dimensions of PWU, and (iii) to propose a more comprehensive framework of user attitudes in the workplace. We sought to achieve our research goals by using inductive, empirically derived techniques of narrative analysis, in particular content analysis, correspondence analysis, and Q-methodology.

RESEARCH METHODS AND RESULTS

Narrative analysis is a widely used tool for producing inductive, but systematically derived results. It enables researchers to use the attitudes of a diverse set of individuals who tell a story in their own words. Data collected in this manner focuses the research on issues that are raised by the participants, without prompting from the researchers. We chose narrative analysis to investigate personal Web usage in the workplace because we were attempting to elicit people's thoughts and feelings on a sensitive issue, and we believed that narratives would yield information not accessible by more traditional methods such as Likert-type response scales (Hoyle et al., 2002). Narrative analysis has been widely used in medical sciences, social sciences, but less frequently in organizational sciences.

In our work, the narrative analysis had two distinct studies. In the first study, we combined content analysis, the dominant technique for narrative analysis, with correspondence analysis. Content analysis is a process by which desired information from the text is systematically extracted and centers on the frequency with which words or themes appear in texts (Babbie, 1995; Jupp & Norris, 1993; Smith, 2000; Weber, 1990). Correspondence analysis builds on content analysis by empirically deriving relationships among these words or themes. The technique also provides insights into the similarities and differences in the content and structure of the different texts (Bendixen, 1996; Carley, 1997; Carley & Palmquist, 1992). In the second study, we examined the importance of the themes by using Q-methodology (McKeown & Thomas, 1988). Q-methodology, created by a British physicist-psychologist, William Stephenson in 1935, involves the rank ordering of a set of statements to explore the positions held by participants (Brown, 1996). It is especially suited for uncovering diverse positions held by participants on sensitive issues rather than accepting categories developed by researchers (Previte, Hearn, & Dann, 2001). The procedures we followed and the results of each study are discussed below.

Study 1

Respondents and Procedures

Two sets of respondents were used in the first study. The first set was part-time MBA students from a leading university in the northeastern United States. Each MBA student provided the name and e-mail address of three other individuals who used the Web at work; this constituted the second set. This "snowballing" data-collection method was consistent with previous work (Stanton & Weiss, 2000) and increased the variability in our sample, a desirable characteristic for inductive research (Hoyle et al., 2002).

We asked everyone to respond electronically to the following open-ended question: *"Do you think it's ok for a person to use the Web for non-work purposes during working hours in the workplace."* We felt that open-ended questions allowed the respondents to answer in a relatively unconstrained way, and that a broad, single question was sufficient to capture the complexities of the phenomenon (Hoyle et al., 2002). This question was the result of a series of pilot tests, in which the wording and clarity were checked.

Since participants typed their responses and sent them electronically, data was gathered verbatim, so there was no possibility of transcription errors, thus enhancing credibility (Corcoran & Stewart, 1998). We also asked for demographic information that included age, gender, education, work experience, and current organizational position.

The high response rate of 89% (481) was attributed to the fact that the participants were either registered in the courses or they were acquainted with the MBA students. Our final sample consisted of 316 responses with complete data, including 110 responses from the first set and 206 from the second set. The majority of the participants were male (67.3%), educated (88% with a bachelor's degree or above), and young (73% reported being between 18 years old to 39). Work experience averaged 16 years, ranging from 1.8 years to 30 years. Managers represented 42% of the participants (top level = 8%; middle level = 14%; and lower level = 20%); professionals represented 32%; and administrative support were 11% of the sample.

Coding the Narratives

The goal of the coding scheme was to capture the major themes and relationships respondents mentioned in their answers. We developed the coding scheme inductively, adding new codes as the respondents mentioned new themes in the different narratives (Haney et al., 1998). The coding process involved five steps and was done by one of the authors and two doctoral students. The use of investigator triangulation, that is using multiple coders, decreases coding bias, thus enhancing objectivity (Kuzel, 1992).

First, based on a preliminary examination of the text, an "initial list" of codes was created. While coding the data, it was noticed that at the beginning of each narrative, the respondents self-categorized themselves regarding their overall perception about personal Web usage at work. An example of this type of categorization was: *"I do not think it's ok to use the Web for personal reasons while at work."* This was followed by a description of their attitudes about PWU. Second, 50 narratives were independently read to develop a list of codes from which 24 themes emerged. Third, these lists were compared, and differences were reconciled, leading to the identification of 19 themes. Fourth, 10 randomly selected narratives were then coded — inter-coder agreement was 75% (Kappa statistic = 0.50). Since the Kappa coefficient was lower than the recommended 0.61 (Kvalseth, 1989), further discussion ensued and another 10 randomly selected narratives were coded. Inter-coder agreement improved to 90% (Kappa = 0.80). Fifth, a coding manual was then developed and used to code the 316 narratives individually. Each narrative was sorted into one of four categories — two categories of respondents who simply expressed approval or disapproval: *'personal Web usage at work is ok'* (YOK); and *'personal Web usage at work is not ok'* (NOK), and two categories with respondent judgments that were qualified: *'personal Web usage at work is ok within limits'* (OKWL); and *'personal Web usage at work is ok as long as productivity doesn't suffer'* (YOKPS). Respondents' answers were then analyzed searching for the 19 themes and dichotomously coding "1" = *theme was mentioned in the text* or "0" = *theme was not mentioned in the text.* Thus narratives could contain more than one theme. The inter-coder agreement was 96% (Kappa statistic = 0.89). Following Krippendorff (1980), disagreements on coding were discussed until agreement was reached.

Data Analysis

The data analysis consisted of three stages: (i) a content analysis, (ii) a correspondence analysis with categories and themes, and (iii) a correspondence analysis with supplementary variables.

In the first stage a content analysis, a simple count of each theme mentioned either explicitly or implicitly by the respondents, was performed. If a respondent mentioned a theme more than once, we counted it as a single mention. This conservative counting rule meant that the total number of mentions in all of the narratives serves as a rough indicator of the relative salience of a theme.

Results — Content Analysis

Table 1 details the coding scheme, showing the four categories, 19 themes, frequencies, and codes. Frequencies in the categories without qualifications, *Yes, PWU is ok* (YOK) and *No, PWU is not ok* (NOK) are almost the same, 65 and 61 respectively. The categories which express qualifications, *Yes, personal access is ok if it doesn't impact productivity* (YOKPS) and *ok only within limits.* (OKWL) are also almost equal with 98 and 92 respondents respectively. The five most frequently mentioned themes were: *"Should have policy"* (SHP), 97; *"Can lead to legal issues"* (LEG), 72; *"Monitoring to*

Table 1. Categories and Theme Frequencies and Definitions

Categories	***f***	**Definitions**
NOK	61	No, PWU is not ok
YOK	65	Yes, PWU is ok
OKWL	92	Ok only within limits, e.g., before working hours
YOKPS	98	Yes, personal access is ok if doesn't impact productivity
Themes		
NMON	7	It's not ok to monitor personal access
CRT	10	Personal usage leads to creativity
BW	16	Bandwidth issues with personal access
RS	17	Personal usage part of required skill sets
LIMA	21	Company should allow limited personal access
PRI	21	Privacy issues with personal access
SCON	25	Soft controls to limit personal access
LPEFFY	27	Personal access leads to loss of productivity and efficiency
TCON	28	Technology-based controls to limit personal access
BT	31	Business tool
POSFE	31	Positive feelings for organization
JTYPE	34	Personal access depends on type of job
WCULT	34	This is the work culture
REX	35	Relaxing
DOO	44	Like doodling or taking a break
PROEFFY	44	Leads to productivity and efficiency
YMON	58	Yes, it's ok to monitor personal access
LEG	72	Legal issues with personal access
SHP	97	Should have a policy

limit personal access" (YMON), 58; *"Like doodling or taking a break"* (DOO), 44; *"Leads to productivity and efficiency"* (PROEFFY), 44.

Then we created a frequency cross-tabulation of the four categories by the 19 themes, shown in Table 2. This table formed the basis for the correspondence analysis, the second stage of our data analysis in Study 1.

In the second stage of our data analysis, we used SPSS v.10 to run a correspondence analysis (CA). The primary goal of this exploratory multivariate statistical technique was to transform each row and each column in the cross-tabulation table into a theme cloud of points with separate points on a map (i.e., the point map). As opposed to traditional hypothesis testing designed to verify *apriori* hypotheses about relationships among variables, CA is used to identify systematic relationships among variables when there are incomplete *apriori* expectations as to the nature of those relationships.

Table 2. Cross-Tabulation Between the Categories and Themes

Themes	YOK	OKWL	NOK	YOKPS	Total
NMON	3	1	1	2	7
CRT	4	2	1	3	10
BW	3	5	4	4	16
RS	6	4	1	6	17
LIMA	2	8	5	6	21
PRI	4	5	5	7	21
SCON	7	5	2	11	25
LPEFFY	1	8	10	8	27
TCON	4	10	2	12	28
BT	10	4	4	13	31
POSFE	9	6	2	14	31
JTYPE	11	7	3	13	34
WCULT	9	10	3	12	34
REX	12	9	2	12	35
DOO	15	14	4	11	44
PROEFFY	18	8	2	16	44
YMON	6	22	9	21	58
LEG	13	22	12	25	72
SHP	16	29	23	29	97
	153	179	95	225	652

Results — Correspondence Analysis with Categories and Themes

The results indicate that there was a significant dependency between the themes and categories ($\chi^2 = 77.38$; df$=54$; $p < 0.05$). A screen plot indicated that a two-dimensional solution was the most suitable, with the first and second principal axes accounting for 76% and 15% of the inertia respectively.

Table 3 provides the dimensions and their correspondence to the categories and themes. The first two numeric columns show the coordinates of the categories and themes of the dimensions. The next two columns provide the contribution to the inertia of the dimensions. The final two columns provide the squared cosine, which is the sum of the squared correlation of a row or column. The final column indicates the total squared cosine values of the two dimensions and is a measure of the quality of representation of each point in the coordinate space (Greenacre, 1984). As can be seen, all categories and themes except for *"like doodling or taking a break"* (0.381) are well represented by the two dimensions.

Table 3. Dimensions and their Correspondence to the Categories and Themes

	Coordinates		Contribributions (%)		Squared cosines		
	F_1	F_2	F_1	F_2	F_1	F_2	Total
Categories							
NOK	-0.518	0.229	**42.848**	**42.406**	0.836	0.164	1.000
YOK	0.425	0.117	**46.390**	17.908	0.904	0.069	0.973
OKWL	-0.173	-.0147	9.028	**32.901**	0.475	0.342	0.817
YOKPS	0.068	-0.060	1.734	6.786	0.182	0.140	0.332
Themes							
NMON	0.340	0.335	1.358	**6.685**	0.494	0.480	0.974
CRT	0.343	0.168	1.984	2.403	0.752	0.180	0.932
BW	-0.288	0.137	2.235	2.568	0.719	0.163	0.882
RS	0.340	-0.005	3.296	0.004	0.984	0.000	0.984
LIMA	-0.429	-0.054	**6.493**	0.528	0.953	0.015	0.969
PRI	-0.202	0.164	1.445	4.813	0.567	0.373	0.941
SCON	0.240	-0.033	2.430	0.231	0.682	0.013	0.695
LPEFFY	-0.687	0.208	**21.400**	4.982	0.904	0.083	0.987
TCON	-0.030	-0.334	0.043	**26.634**	0.008	0.941	0.949
BT	0.252	0.175	3.318	**8.052**	0.497	0.239	0.736
POSFE	0.288	-0.048	4.318	0.619	0.719	0.020	0.739
JTYPE	0.271	0.038	4.206	0.423	0.949	0.019	0.968
WCULT	0.131	-0.097	0.986	2.706	0.614	0.333	0.947
REX	0.313	-0.037	**5.783**	0.397	0.937	0.013	0.950
DOO	0.197	-0.006	2.872	0.015	0.381	0.000	***0.381***
PROEFFY	0.474	0.075	**16.658**	2.085	0.975	0.024	0.999
YMON	-0.257	-0.221	**6.442**	**24.036**	0.575	0.424	1.000
LEG	-0.129	-0.046	2.025	1.319	0.880	0.113	0.993
SHP	-0.279	0.089	**12.707**	**5.501**	0.902	0.091	0.993

Figure 1 illustrates the spatial association of the theme and category clouds of points, as defined by the two principal axes. The plots were merged into one joint display through a canonical normalization procedure. This allowed the proper interpretation of distances between any row items and the distance between column items, as well as the distance among row and column items (Greenacre, 1993). The axes were interpreted by way of the contribution that each point made towards the total inertia. In this study there were 19 perceptual themes, and any contribution greater than 5.26% (i.e., 100%/19) would indicate a significance greater than what would be expected in the case of a purely random distribution of themes over the axes (Greenacre, 1993).

Dimension 1 (76%): On the positive side of this dimension, we found two categories of responses: *Yes, PWU is ok* (YOK) and *Yes, personal access is ok if it doesn't impact productivity* (YOKPS). On the negative side we find *No, PWU is not ok* (NOK) and *ok only within limits.* (OKWL) The contributions indicate that the categories that have the most impact in determining the orientation of this dimension were YOK, with 46.3% of the inertia, anchoring the positive end, and NOK with 42.8% of the inertia, anchoring the negative end.

For interpretation of this dimension, we turn to the coordinates and contributions of the perceptual themes. The contribution to inertia of the perceptual themes indicates that the first principal axis is determined by:

- two themes with positive coordinates: leads to productivity (PROEFFY), 16.6%; and relaxing (REX), 5.7%; and
- four themes with negative coordinates: loss of productivity and efficiency (LPEFFY), 21.4%; should have policies (SHP), 12.7%; yes, monitoring is ok (YMON), 6.4%; and company should allow within limits (LIMA),

Figure 1. Themes and Dimensions

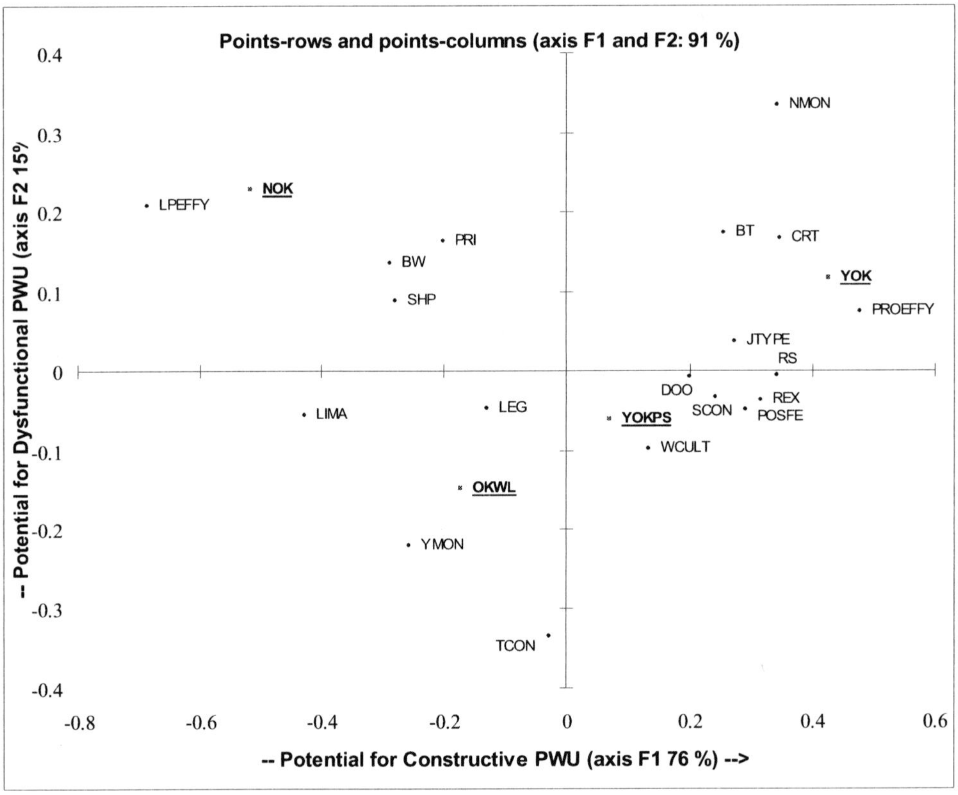

6.4%. Based on these themes, we interpret this as a distinction between high and low potential for constructive personal Web usage and label this dimension *"Potential for constructive personal Web usage."*

Dimension 2 (15%): Categories NOK 42.4% and YOK 17.9% have high positive scores on the contributions to inertia. OKWL 32.9% and YOKPS 6.7% were on the negative side of this dimension. The second principal axis was determined by the following themes: business tool (BT), 8.0%; no, it's not ok to monitor (NMON), 6.6%; loss of productivity and efficiency (LPEFFY), 5.9%; and should have policies (SHP), 5.5%. All of these themes had positive coordinates. Technical controls (TCON), 26.6%; and yes, monitoring is ok (YMON), 24% were the themes which had negative coordinates. Based on the largest positive and negative coordinates, the second dimension was labeled *"Potential for dysfunctional personal Web usage."*

In the third stage, we did a correspondence analysis where the supplementary variables of age, gender, education, experience, and current organizational position were projected into the theme/category space developed in Stage 2. Since these variables were projected after the construction of the factorial axes in the new axes set, these supplementary points had a position in the full space, but did not affect the positioning of the theme points.

Results — Correspondence Analysis with Supplementary Variables

Of the supplementary variables only current organizational position had a cosine that was high enough to warrant its inclusion in the two-dimensional solution (Greenacre, 1984). Figure 2 shows attitudes of the potential dysfunctional or constructive nature of PWU vary by organizational position. Top-level managers' attitudes group together in the middle of the map, indicating they perceived personal Web usage in the workplace as moderately dysfunctional as well as moderately constructive. Middle-level managers' responses are positioned in the lower-right quadrant, seeing PWU as having higher constructive potential and lower dysfunctional potential. Lower-level managers' comments are clustered in the upper-right quadrant, perceiving PWU's potential for both dysfunctional and constructive usage as high. Professionals report that PWU has moderate potential for abuse, coupled with higher constructive potential. The comments of respondents with administrative positions are in the lower-left quadrant, viewing PWU as having moderate dysfunctional potential with low constructive potential.

Figure 2. Themes, Dimensions, and Organizational Positions

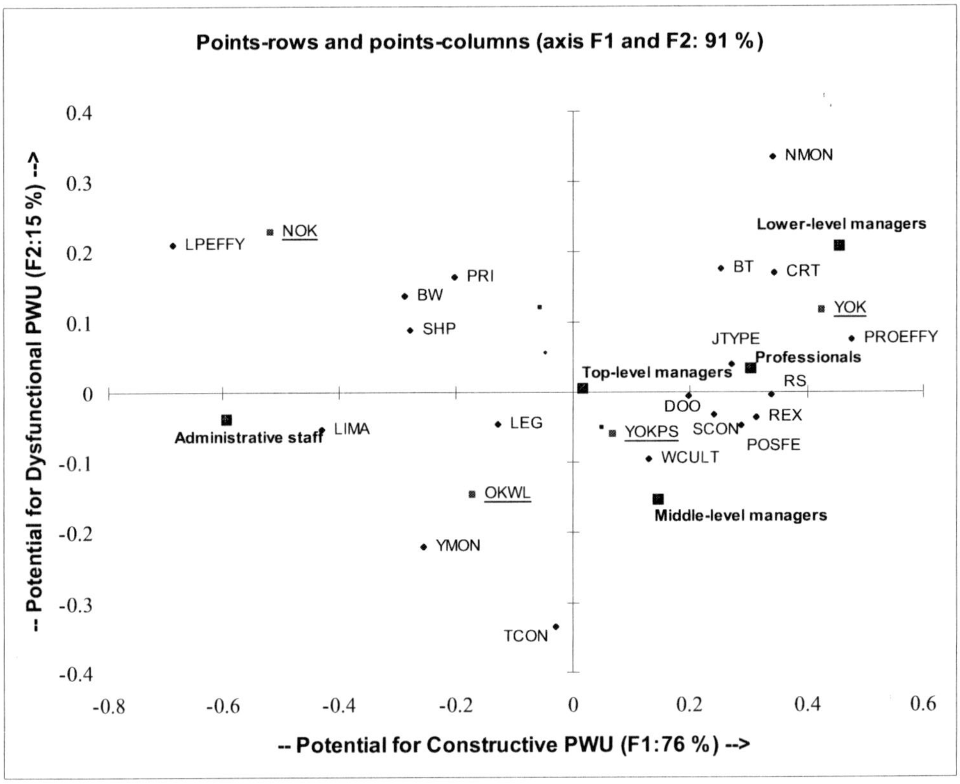

Study 2

We used Q-methodology to examine the consensus viewpoints of respondent attitudes on personal Web usage behavior in the workplace to extend our Study 1 findings. This type of small sample analysis is useful in profiling attitudes about a phenomenon, seeking to measure the relative importance of personal beliefs on issues or debates of social or economic consequence (Addams & Proops, 2000; Brown et al., 1999; Carlson & Williams, 1993). Q-methodology has enjoyed a long history of acceptance and use in political science, journalism, and psychology (Brown, 1968), while its use in business research has been rather limited (Chatman, 1989, 1991; Kleine, Kleine, & Allen, 1995). It is important to note that Q-methodology highlights the assortment and type of viewpoints, but not the proportion of a population that holds certain viewpoints (Carlson & Williams, 1993).

Respondents and Procedures

The initial 315 respondents were contacted to build a small convenient sample. Q-methodology is an intensive approach that focuses on the attitudes of a few people using many questions, rather than the reactions of a large number of people to a smaller number of questions. This small sample technique provides depth rather than generalizability and is particularly appropriate for sensitive topic research (Carleson & Williams, 1993). We used a sample of 25 participants, representative of the five organizational positions (top-level managers = 4; middle-level managers = 5; lower-level managers = 4; professionals = 3; and administrators = 9). These participants were given 38 statements derived from the narratives representing the 19 themes. These statements reflected personal Web usage behaviors. Examples of these statements are:

- *"Because employees are working longer hours, they need to log on to the Web during work hours for personal reasons."*
- *"Certain types of websites should be blocked by the organization."*
- *"Personal Web usage offers opportunities to promote employee creativity."*

Participants evaluated the statements along a continuum ranging from "-4" (strongly disagree) to "+4" (strongly agree). The forced choice format of the Q-sort process made the results fall into a quasi-normal distribution (Carlson & Williams, 1993). The respondents used a Web-based Q-sort methodology, allowing seamless data entry and recording.

Data Analysis

The completed Q-sorts were analyzed using an inverted factor analysis technique. In this technique, interpretations are based on factor arrays and factor scores rather than loadings, typically used in factor analysis. Thus, groups were formed based on common viewpoints. In Table 4 the correlations between the participants and factors are given. A three-factor solution emerged. In Factor 1 were eight of the nine administrations, three of four top-level managers, and two out of the four middle-level managers (total 13). In Factor 2 were all three of the professionals and one lower- and one middle-level manager (total 5). In Factor 3 were three of the four lower-level managers and one middle-level manager (total 4). Three respondents loaded on two factors.

Table 4. Correlations Between Participants and Factors

Respondent	Organizational Position	Factors I	II	III
1	Administrator	0.562 *	-0.550	0.059
2	Administrator	0.799 *	0.059	0.028
3	Professional	0.058	0.880 *	-0.147
4	Lower-level Manager	-0.033	0.158	0.986 *
5	Administrator	0.740 *	0.034	0.011
6	Top-level Manager	0.548 *	0.554 *	-0.042
7	Top-level Manager	0.646 *	-0.426	0.057
8	Middle-level Manager	0.697 *	0.059	0.125
9	Professional	0.036	0.780 *	-0.199
10	Lower-level Manager	-0.443	0.058	0.909 *
11	Administrator	0.623 *	0.036	-0.011
12	Middle-level Manager	0.571 *	0.550 *	-0.044
13	Middle-level Manager	0.622 *	-0.496	0.059
14	Administrator	0.799 *	0.059	0.028
15	Middle-level Manager	0.050	0.704 *	-0.169
16	Lower-level Manager	-0.065	0.200	0.809 *
17	Administrator	0.735 *	0.026	0.018
18	Top-level Manager	0.577 *	0.538	-0.024
19	Administrator	0.663 *	-0.596 *	0.059
20	Administrator	0.799 *	0.059	0.038
21	Lower-level Manager	0.090	0.810 *	-0.179
22	Middle-level Manager	-0.033	0.180	-0.861 *
23	Administrator	0.782 *	0.002	0.017
24	Top-level Manager	0.648 *	0.520	-0.045
25	Professional	0.099	0.765 *	-0.191
Expl.Var		7.231	6.430	4.022
Prp.Totl		0.289	0.257	0.161

* Significant Loadings p< 0.05

The next step, shown in Table 5, was to label the factors based on the factor arrays. Particularly important for labeling are the statements at the extremes, i.e., *most agree* (+4) to *most disagree* (-4). It is sufficient to analyze the rounded scores (+4 to -4) since as a general rule, differences in scores of two or more are considered significant at the $p<0.01$ level (Addams & Proops, 2000). The table of factor scores indicates the extent to which each of the 38 statements characterizes each of the three factors. The statements associated with each factor are discussed in the following section.

Table 5. Factor Arrays of Personal Web Usage Profiles

Statement #		Profiles A	Profiles B	Profiles C
	Profile A: Cyber-Bureaucrat			
4	Certain types of websites should be blocked by the organization	+3	-2	-4
6	Companies should control personal web usage at work by blocking out as much objectionable material as possible via firewalls	+4	-2	-4
14	If employees are downloading very large files, like music files simultaneously, there are financial implications because bandwidths could be overloaded	+4	+1	-1
16	Personal web usage can expose the organization to risks such as viruses	+3	+1	-2
21	The cost of bandwidth could be an issue for organizations with employees using the web for personal use	+3	+1	0
22	Personal web usage at work should be discouraged because it limits work efficiency	+4	-4	-2
24	The employee should be mindful of hardware and software constraints and costs and not download files that could burden the company's resources	+3	+2	0
25	The organization should generate reports on employees' personal web usage behavior	+3	-3	-1
26	The organization should monitor employee's personal usage of the web	+4	-3	+1
27	Through personal web usage, a company's name could be dropped into questionable sites potentially tarnishing a company's name	+3	+1	+1
28	Unmonitored personal web usage at work can lead to abuse	+4	+2	+1
30	Using the web for nonwork purposes is a time consumption that can adversely affect work production	+4	-4	-1
31	Web usage policies are important in controlling personal web usage	+3	+2	+1
32	When an employee uses the web for personal reasons they invade the privacy of other workers	+3	+1	-2
34	With unmonitored use of the web, employees could be offended if other employees view pornography sites are viewed during working hours	+4	-2	-2
35	Your company can actually benefit from your surfing the net for non-work purposes	-4	+2	+2
36	A company should have clear and precise web policies outlining do's and don'ts on personal web usage	+3	+2	0
	Profile B: Cyber-Humanist			
2	Because employees are working longer hours they need to log on to the web during work hours for personal reasons	-3	+4	+2
3	Breaks such as smoke and coffee breaks are similar to using the web for personal reasons	-2	+4	+3
5	Companies might allow personal web usage as a way to reward employees for the purpose of improving morale	-1	+4	+2

Table 5. Factor Arrays of Personal Web Usage Profiles (continued)

Statement #		Profiles A	Profiles B	Profiles C
8	Doing something mindless like checking the latest college ranking on the web can help an employee relax during working hours	-2	+3	+2
9	Employers should trust their employees not to abuse their personal web usage privileges	-3	+4	+3
10	I believe nonwork use of the web should be treated like phone usage	-4	+2	+3
12	I think that employees would be happier with their jobs if they had could use the Web for personal reasons during working hours	-1	+3	+1
13	I would be more productive if my employer allowed me to use the web for personal reasons during working hours	-3	+4	+1
15	If someone told me not to use the web personal reasons during work hours, it would change my attitude towards the company	-2	+3	+1
17	Personal web usage can relieve an employee's stress during the workday	-1	+3	+1
18	Personal web usage gives me a break from having to concentrate all day long	-2	+4	+1
19	Personal web usage leads to a more productive worker	-4	+4	+3
33	With the long hours spent at one's work, a little taste of freedom by using the web for personal use is a definite morale lifter	-2	+4	+2
37	Personal web usage can enhance and improve the function of the employees	-2	+3	+2
	Profile C: Cyber-Adventurer			
1	An employee engaged in self-enrichment, training, and knowledge increasing via the web is able to perform better than their counterparts	-2	+2	+4
7	Companies should encourage employees to surf the web to look for ways to increase performance and keep on top of financial news	-3	+2	+3
11	I can work on developing myself, looking for new things that improve knowledge and skills through personal web usage	-3	+4	+3
20	Personal web usage offers opportunities to promote employee creativity	-4	+2	+4
23	Personal web usage can help employees to be better educated and knowledgeable about the business environment	-2	+2	+4
29	Users should be given discretion to use the web for personal reasons	-4	+3	+4
38	Some employees' jobs require them to use the web for personal reasons	-2	+1	+4

Results

Profile A (Factor 1) — Cyber-Bureaucrat: This profile illustrates attitudes of Web users who feel that PWU should not occur during working hours. The mind-set is that PWU during working hours leads to inefficiency (Statements #22, 30). These Web users perceive that usage leads to a plethora of challenges such as: PWU causes clogging of the networks (Statements #14, 21, 24), a higher likelihood of security and privacy concerns (Statements #16, 32), and possible legal problems (Statements #27, 34). In addition these people perceive that PWU should be controlled by technical controls (Statements #4, 6), clearly stated policies (Statements #31, 36), and by monitoring (Statements #25, 26, 28).

Overall the focus of this profile, primarily administrators, top-level managers, and half of the middle-level managers, seems to be that PWU had little value primarily because it was too difficult to monitor usage. They felt that unsupervised and unmonitored personal usage put the company at risk and that a systematic and objective system of managing Web usage was needed. This profile is compatible with a scientific, bureaucratic view of work where hierarchy, controls, formal communication, and written policies and procedures define the workplace.

Profile B (Factor 2) — Cyber-Humanist: Responses in this profile exhibit generally positive attitudes towards personal Web usage at the workplace. For instance respondents in this profile believe that there is a need to balance working and living, and that the Internet-connected workplace has made the lines fuzzy between work and non-work (Statements #2, 33). These people perceive PWU as having positive affective outcomes (Statements #5, 12, 15). The positive feelings about the workplace lead to the potential for more constructive Web usage, which spills over to higher productivity in the workplace (Statements #13, 19). PWU is seen as equivalent to taking a break (Statement #3) and is relaxing (Statements #8, 17). They feel *employers should trust their employees not to abuse their personal Web usage privileges* (Statement #9).

This profile is closely aligned with the social and psychological needs of employees, and is "people and relationship" oriented. Responses in this grouping reflect a growing body of research that associates organizational success with treating employees as assets, rather than costs (Pfeffer & Veiga, 1999). Employees are to be considered trusted partners who can self-regulate their behaviors. It is not surprising that this profile consists mainly of the professionals since they are often associated with self-motivation and self-governance.

Profile C (Factor 3) — Cyber-Adventurer: In the third profile, a common view is that *'Users should be given discretion to use the Web for personal reasons'* (Statement #29). The "rules of the game" are not established and the adventurers fill the void by creating rules and adapting situations for their advantage as the following statement illustrates: *Companies should encourage employees to surf the Web to look for ways to increase performance* (Statement #11). There is potential for constructive usage, *'Personal Web usage can help employees to be better educated and knowledgeable about the business environment'* (Statement #23). *PWU is seen as promoting creativity* (Statement #20) and *improving knowledge and skills* (Statement #11). Cyber-adventurers view PWU as a way to engage in self-enrichment and training to improve performance (Statements #1, 23).

The cyber-adventurer can be described as exhibiting individualistic or entrepreneurial-like attitudes. There is optimism that the Web will place her/him on the frontier of continuous self-improvement, a goal worth the risk of potentially dysfunctional outcomes. This profile primarily consisted of responses from lower-level managers who might be most open to taking risks to improve their positioning.

DISCUSSION

Our goal was to empirically research the issue of personal Web usage in the workplace by mapping this concept from the vantage point of employee attitudes. We suspected that personal Web usage in the workplace was a complex issue, with the potential for dysfunctional behaviors as well as constructive behaviors. This interest in both the potential for positive and negative consequences of personal Web usage was a departure from previous work on personal Web usage that focused almost exclusively on the negative effects (Joinson, 1998; Griffiths, 1998; Putnam & Maheu, 2000) or that posited that personal Web usage was just another way of wasting time at work (Block, 2001). We also made a contribution to the literature by using qualitative methodology in contrast to survey data and regression analyses, building on the work of Klein and Meyers (1999).

Our study produces a more comprehensive framework of personal Web usage at work. We identify 19 themes and four categories using responses to the question, *"Do you think it's ok for a person to use the Web for non-work purposes during working hours in the workplace?"* Through correspondence analysis, we identify systematic relationships between the themes, with a two-dimensional solution best fitting the data. The first dimension we named *potential for constructive Web usage* and the second *potential for dysfunctional personal Web usage*. Third, we take our study an additional step by overlaying employee position onto the first two analyses, and we discover that job positions are uniquely placed. Top-level managers' attitudes group together in the middle of the map, falling in between the two clusters, perceiving both moderate dysfunctional and constructive potential. This may be indicative of the propensity of top managers to look at issues from multiple perspectives, reflecting their need to consider multiple stakeholders, both internal and external to the organization. Middle managers are in the lower-right quadrant, perceiving higher constructive potential and lower dysfunctional potential. Professionals see moderate potential for abuse, with higher constructive potential. The proximity of these two groups is consistent with their interpersonal focus and intermediator's role between top and lower management. They mix in rationalizations such as personal Web usage at work is ok to do their jobs better, and address the increasing spillover of work into non-work time. Lower-level managers are in the upper-right quadrant, representing the highest potential for both dysfunctional and constructive usage. Perhaps this group feels the strongest needs to escape the pressures of managing and to build skill sets for upward mobility. The respondents with administrative positions perceive moderate dysfunctional potential with low constructive potential, a result consistent with their focus on efficiency and transactions.

Finally, we used Q-methodology to further extend our investigation, allowing us to profile users' attitudes towards PWU behaviors. Using the 19 themes we created a list of 38 behaviors (two for each theme). This analysis resulted in three profiles, which we named cyber-bureaucrat, cyber-humanist, and cyber-adventurer. These profiles were generally consistent with the correspondence analysis findings, indicating the critical nature of employee position in attitudes towards PWU. The importance of employee position and lack of significance of the other demographic variables was unanticipated.

Based on our results, another major contribution of our work is to construct a definition of personal Web usage in the workplace that is grounded in empirical analysis. We would like to suggest the following definition: *"voluntary online Web behaviors during working time using any of the organization's resources for activities outside current customary job/work requirements."* We limited the definition to "online behaviors" to separate it from other behaviors such as listening to music or playing games already downloaded. We broaden the time to "working time" to allow for work done off premises and/or outside of the normal nine to five office hours. Surfing suggests aimless access, but much of personal Web usage has specific destinations or purposes such as travel arrangements or personal websites, so we deleted the word "surfing." We also wanted to indicate the use not only of the company's network and servers, but also the use of computers and employees' time, hence the wording: "organization's resources." Finally, we wanted to indicate that the personal Web usage was outside of current customary job/work requirements, suggesting a potential for learning not associated with existing job/work requirements.

Another contribution of our present study is that it should prove useful in extending the social contract theory to the 21st century work environment. The social contract theory suggests that humans evolve ways of dealing with other humans, with groups, and within organizations by establishing commonly accepted rules of conduct (Cosmides & Tooby, 1992). In the past, two types of social contracts defined the work environment. There is an *economic contract* where wages, fringe benefits, and reasonable working conditions are exchanged for time, skills, and effort. There is also a *psychological contract* where a certain amount of allegiance, creativity, and extra effort are exchanged for job security, fair treatment, rewarding relationships with coworkers, and organizational support (Shore & Tetrick, 1994).

The nature of the employment relationship is shifting, and the mechanisms controlling these relationships are no longer clear. As the work environment becomes more flexible, open, and autonomous, and the work becomes more disassociated with a specific brick-and-mortar place and specific job requirements, the exchange mechanisms and processes become less certain. The line between work and life becomes fuzzy, and work is defined as 24x7 (24 hours a day, seven days a week). As one respondent stated:

> *In the corporate world, there is no longer such a thing as "the 40-hour work week," nor do people work from "nine to five." Today, people are required to work all sorts of different shifts and the average work day is probably closer to nine or 10 hours, depending on your position with the company. That being said, I think it is wise for a corporation to allow its employees to utilize the Internet for their personal uses. It gives the employee some time to themselves, where they can just "veg-out" for a few minutes or actually do something constructive, like online banking.*

Thus, the concepts of the psychological contracts and jobs as we know them may no longer be valid (Bridge, 1995). Lim, Teo, and Loo (2002) report individuals rationalize that they are entitled to spend time on the Web on personal issues while at work as a form of informal compensation. This is consistent with our findings; a respondent succinctly states:

> *I am salaried and very often required to work long hours, have working lunches (not taking a client out socially, but listening to venders sell their financial products), go to out-of-town meetings and seminars that take up an entire weekend, etc. If I happen to have a free hour on a Tuesday morning or two hours on an occasional afternoon and I choose to use the Internet, my computer, or work on my MBA online, then I truly believe it is ok, because my company is getting back many more hours of my time. My schedule would be considered flexible. Yes, I probably should do it at home, but the fact is, I might have to stay for four more hours due to some other work commitment and I can't very well drive home and back (two hours) to spend the two hours on the Internet.*

In the 21st century work environment, the emphasis on a knowledge workforce is increasing (Brynjolfsson, 1993; Johannessen et al., 2001). The Web can be used to expand the total knowledge base — the tacit, the explicit, the internal, and the external for both the individual and the organization (Dewett & Jones, 2001; Johannessen et al., 2001; Powell & Dent-Micallef, 1997). One of our respondents concisely describes this constructive dimension of PWU:

> *I think that it is alright to use the Web for non-work purposes during work hours. I use the Web at work as a source of information to keep me up to date with current events. Through the Web, I can follow the latest business news as well as world events. I believe that by staying on top of current business news that I become a better-informed knowledge worker.*

Hence, we posit that organizational mechanisms for controlling the social exchange in an Internet-connect work environment may be expanding to include a cyber-contract, which describes the exchange mechanisms in a Web-connected workplace These mechanisms are based on principles, not rules and standards, and are negotiated, openly communicated, and flexible, to adjust for continuous work, learning, and change.

We simply have not had enough experience with this type of exchange process to know what is dysfunctional and what is constructive behavior and how best to manage it for the mutual benefit of individual and organization. The more abstract the entities involvement and the more abstract the work environment, the fewer effective mechanisms we have for control of social exchanges (Allen, 1999). The dangers of the undesirable dysfunctional outcomes of PWU such as loss of intellectual property, sexual harassment, and security risks are real and have led to organizations controlling PWU. We suggest from our study, too much freedom can be dysfunctional, that is, leading to *cyber-slacking*. However, there is also a danger that too much control of personal Web usage can be a danger by stifling creativity, learning, and positive job feelings, leading to what we have termed *cyber-silencing*. We suggest that a middle ground is evolving between unrestricted access and too many restrictions — what we have called *cyber-stimulating*. In this zone the aim is to stimulate learning, leading to a productive usage of the Web in the workplace while isolating dysfunctional and threatening usage.

From our work, we mapped the profiles onto the two-dimensional space, depicting a model of personal Web usage in the workplace (Figure 3). In this model, we posited that personal Web usage in the workplace is a range or area where there is a balance between too much and too little control and is the bottom-right quadrant. This zone of stimulation should balance complexity reduction through too much control and complexity absorption through too much freedom (Boisot, 1999). It should allow for change and growth for both employees and organizations. The upper-left and upper-right quadrants represent cyber-slacking, where PWU has the potential to degenerate into dysfunctional behaviors. The bottom-left quadrant represents cyber-silencing where, while the potential for dysfunctional behavior is lessened, so too is the potential for constructive outcomes from the personal usage of the Web.

The contributions of our work are limited by several factors. Despite steps taken to control coding bias, the interpretation and coding of the responses is subjective in nature. The study is also limited in its generalizability due to the use of convenience sampling rather than random sampling. Additionally, a methodological limitation of this study is the arch effect in the correspondence analysis. This is a typical artifact in an ordination diagram, in which the second axis is an arched function of the first axis. Future studies should attempt to use a detrended correspondence analysis technique.

Figure 3. Personal Web Usage in the Workplace Model

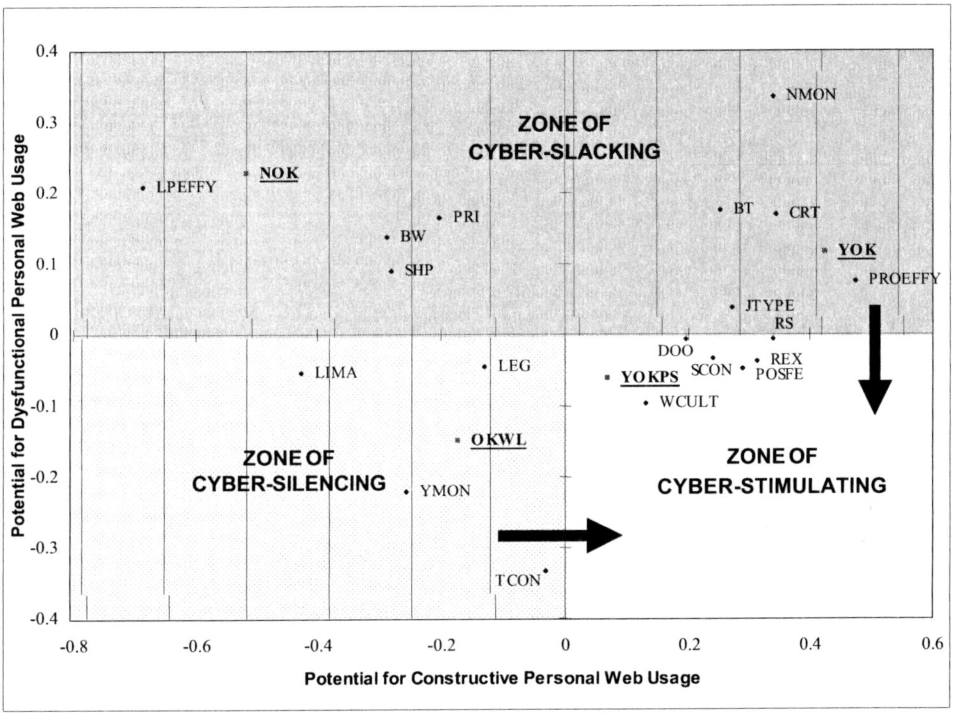

FUTURE RESEARCH

Research should continue on the construct of personal Web usage to delineate specific behaviors, perhaps using multi-dimensional scaling, to empirically confirm our concept maps, particularly the two-dimension solution of both constructive and dysfunctional roles of PWU. Models of PWU with antecedents and outcomes need to be developed and tested. These models might include individual, group, and organizational variables. The extension of the social contract theory and the model of PWU (Figure 3) need to be empirically studied for verification and modification. There are also important human resource issues such as promotion, discipline, and career pathing that can be linked to this model. The work on job position and profiles are promising lines of inquiry for further exploration.

An interesting question is if the individual profiles we have identified can be extended to profiles of organizations and if these organizational profiles can be mapped onto a similar model as the one shown in Figure 3. There is a need for research that examines the implications of PWU for organizational strategy and whether cyber-stimulating moderates the effects of strategy on organizational outcomes such as innovation, learning, or performance. Another major extension of this research is to examine the influence of national culture on PWU.

We hope that our study will bring attention to the interplay between freedom and control in the Web-connected workplace. It is our intention that our work serves as a catalyst for additional theoretical and empirical research into PWU in the workplace, and how beneficial and detrimental dimensions dynamically interact in defining our 21st century workplaces.

REFERENCES

Addams, H. & Proops, J. (2000). *Social Discourse and Environmental Policy: An Application of Q Methodology.* Cheltenham, UK: Edward Elgar.

Allen, G. (1999). *Software piracy: Why honest people cheat.* Unpublished Paper. Academy of Management, Annual Meeting.

Anandarajan, M., Simmers, C., & Igbaria, M. (2000). An exploratory investigation of the antecedents and impact of internet usage: An individual perspective. *Behavior & Information Technology,* 19(1), 69-85.

Armstrong, L., Phillips, J. G., & Saling, L. L. (2000). Potential determinants of heavier Internet usage. *International Journal of Human-Computer Studies,* (53), 537-550.

Babbie, E. (1995). *The Practice of Social Research,* (7th ed). New York: Wadsworth Publishing Co.

Bendixen, M. (1996). A practical guide to the use of correspondence analysis in marketing research. *Marketing Research On-Line,* (1), 16-38.

Bentler, P.M. & Speckart, G. (n.d.). Attitudes cause behaviors: A structural equations analysis. *Journal of Personality and Social Psychology,* (40), 226-238.

Block, W. (2001). Cyberslacking, business ethics and managerial economics. *Journal of Business Ethics,* 33(3), 225-231.

Boisot, M. (1999). *Knowledge Assets.* Oxford, UK: Oxford University Press.

Bridge, W. (1995). *JobShift.* Reading, MA: Addison-Wesley.

Brown, S. R. (1968). Bibliography on Q technique and its methodology. *Perceptual and Motor Skills,* (April), 587-613.

Brown, S. R. (1996). Q methodology and qualitative research. *Qualitative Health Research,* 6(4), 561-567.

Brown, S. R., Durning D. W., & Selden, S. C. (1999). Q methodology. In G. J. Miller & M. L. Whicker (Eds.), *Handbook of Research Methods in Public Administration*, (pp. 599-637). New York: Marcel Dekker.

Brynjolfsson, E. (1993). The productivity paradox of information technology. *Communications of ACM,* 36(12), 67-77.

Carley, K. (1997). Coding choices for textual analysis: A comparison of content analysis and map analysis. In C. W. Roberts (Ed.), *Text Analysis for the Social Sciences*. Mahweh, NJ: Lawrence Erlbaum Associates Publishers.

Carley, K. & Palmquist. (1992). Extracting, representing, and analyzing mental models. *Social Forces,* 70(3), 601-636.

Carlson, J. M. & Williams, T. (1993). Perspectives on the seriousness of crimes. *Social Science Research,* (22), 190-207.

Chatman, J. A. (1989). Improving interactional organizational research: A model of person-organization fit. *Academy of Management Review*, (July), 333-349.

Chatman, J. A. (1991). Matching people and organizations: Selection and socialization in public accounting firms. *Administrative Science Quarterly*, (September), 450-484.

Conlin, M. (2000). Workers, surf at your own risk. *Business Week*, (June), 105-106.

Corcoran, J. A. & Stewart, M. (1998). Stories of stuttering: A qualitative analysis of interview narratives. *Journal of Fluency Disorders,* 23(4), 247-264.

Cosmides, L. & Tooby, J. (1992). Cognitive adaptations for social exchange. In J. H. Barkow, L. Cosmedes, & J. Tooby (Eds.), *The Adapted Mind.* New York: Oxford University Press.

Dewett, T. & Jones, G. (2001). The role of information technology in the organization: A review, model, and assessment. *Journal of Management,* (27), 313-346.

Friedman, W. H. (2000). Is the answer to Internet addiction, Internet interdiction? In H. M. Chung (Ed.), *Proceedings of the 2000 Americas Conference on Information Systems*. AMCIS.

Greenacre, M. J. (1984). *Theory and Application of Correspondence Analysis.* London: Academic Press.

Greengard, S. (2000). The high cost of cyber slacking. *Workforce,* 79(12), 22-24.

Griffiths, M. (1998). Internet addiction: Does it really exist? In J. Gackenbach (Ed.), *Psychology and the Internet: Intrapersonal, Interpersonal, and Transpersonal Implications.* San Diego, CA: Academic Press.

Haney, W., Russell, M., Gulek, C., & Fierros, E. (1998). Drawing on education: Using student drawings to promote middle school improvement. *Schools in the Middle,* 7(3), 38-43.

Hoyle, R. H, Harris, M. J., & Judd, C. M. (2002). *Research Methods in Social Relations*, (7th ed.). Wadsworth Thomson Learning.

Johannessen, J. A., Olaisen, J., & Olsen, B. (2001). Mismanagement of tacit knowledge: The importance of tacit knowledge, the danger of information technology, and what to do about it. *International Journal of Information Management,* (21), 3-20.

Joinson, A. (1998). Causes and implications of disinhibited behavior on the Internet. In J. Gackenbach (Ed.), *Psychology and the Internet: Intrapersonal, Interpersonal, and Transpersonal Implications.* San Diego, CA: Academic Press.

Jupp, V. & Norris, C. (1993). Traditions in documentary analysis. In M. Hammersley (Ed.), *Social Research: Philosophy, Politics and Practice.* UK: Open University Press.

Klein, H. & Meyers, M. (1999). A set of principles for conducting and evaluating interpretive field studies in information systems. *MIS Quarterly*, (March).

Kleine, S., Schultz, R., Kleine III, E., & Allen, C. T. (1995). How is a possession 'Me' or 'Not Me'?: Characteristic types and an antecedent of material possession attachment. *Journal of Consumer Behavior,* (December), 327-343.

Korgaonkar, P. K. & Wolin, L. D. (1999). A multivariate analysis of Web usage. *Journal of Advertising Research,* 39(2), 53-68.

Krippendorff, K. (1980). *Content Analysis: An Introduction to its Methodology.* London: Sage.

Kuzel, A. J. (1992). Sampling in qualitative inquiry. In B. F. Crabtree & W. L. Miller (Eds.), *Doing Qualitative Research.* Newbury Park, CA: Sage.

Kvalseth, T. O. (1989). Note on Cohen's Kappa. *Psychological Reports* (65), 223-226.

Lederer, A. L., Maupin, D. J., Sena, M. P., & Zhuang, Y. (2000). The technology acceptance model and the World Wide Web. *Decision Support Systems,* 29(3), 269-282.

Lim, V. K. G., Teo, T. S. H., & Loo, G. L. (2002). How do I load here? Let me count the way. *Communications of the ACM,* 45(1), 66-70.

McKeown, B. & Thomas, D. (1998). *Q Methodology.* Newbury Park, CA: Sage.

McWilliams, G. & Stepanek, M. (1998). *Taming the info monster. Business Week*, (June), 170-172.

Pfeffer, J. & Veiga, J.F. (1999). Putting people first for organizational success. *Academy of Management Executive,* 13(2), 37-48.

Powell, T.C. & Dent-Micallef, A. (1997). Information technology as competitive advantage: The role of human, business, and technology resources. *Strategic Management Journal,* 18(5), 375-405.

Putnam, D. E. & Maheu, M. M. (2000). Online sexual addiction and compulsivity: Integrating web resources and behavioral telehealth in treatment. *Sexual Addiction & Compulsivity* (7), 91-112.

Shore, L. M. & Tetrick, A. (1994). The psychological contract as an explanatory framework in the employment relationship. C. L. Cooper & D. M. Rousseau (Eds.), *Trends in Organizational Behavior*, (pp. 58-70). London: John Wiley & Sons.

Simmers, C. A. (2002). Aligning Internet usage with business priorities. *Communications of the ACM,* 45(1), 1-4.

Smith, C. P. (2000). Content analysis and narrative analysis. In H. T. Reis & C. M. Judd (Eds.), *Handbook of Research Methods in Social and Personality Psychology,* (pp. 313-335). Cambridge, MA: Cambridge University Press.

Stanton, J. M. & Weiss, E. M. (2000). Electronic monitoring in their own words: An exploratory study of employees' experiences with new types of surveillance. *Computers in Human Behavior,* (16), 423-440.

Stewart, F. (2000). Internet acceptable use policies: Navigating the management, legal, and technical issues. *Security Management*, (July/August), 46-52.

Sunoo, B.P. (1996). The employee may be loafing. *Personnel Journal,* (December), 55-2.

Teo, T. S. H. & Lim, V. K. G. (1998). Usage and perceptions of the Internet: What has age got to do with it? *Cyberpsychology & Behavior,* 1(4), 371-381.

Teo, T. S. H., Lim, V. K. G., & Lai, R. Y. C. (1999). Intrinsic and extrinsic motivation in Internet usage. *Omega,* (27), 25-37.

Terr, L. (1999). *Beyond Love and Work: Why Adults Need to Play.* Scribner, p. 226.

Weber, R. P. (1990). *Basic content analysis*, (2nd ed.). London: Sage Publications.

Chapter II

Personal Web Usage in Organizations

Zoonky Lee
University of Nebraska - Lincoln, USA

Younghwa Lee
University of Colorado at Boulder, USA

Yongbeom Kim
Fairleigh Dickinson University, USA

ABSTRACT

This chapter presents an empirical investigation of why employees use the Internet for personal purpose during work hours. We are especially interested in perceptual difference between personal Web usage groups and non-personal Web usage groups in the context of non-work-related usage of the Internet. Drawing from previous studies in behavioral intention and human attitude, criminology, and moral and ethical decision-making, a comprehensive model was developed and tested through a field survey of 546 business professionals.

INTRODUCTION

Personal Web usage, non-work related use of the Internet during work hours, is a pervasive behavior observed in our daily work environment. It is reported that 37% of working hours are devoted to personal Web usage (Conlin, 2000). As the cost associated with it is estimated to be $54 billion annually, companies are seeking methods to reduce it. Enforcing Internet use policies and installing filtering or monitoring systems (e.g., Cyber Patrol) are popular trends. An Internet usage survey noted that 68% of the surveyed companies have Internet usage policies (Infoworld, 2000), that 31% of U.S. companies have spent money on Internet-based activity monitoring and filtering systems (Hancock, 1999). Personal Web usage, however, does not seem to be declining but instead is spreading throughout the workplace. We need to understand why this trend has become pervasive despite organizational efforts to reduce it.

In today's computer-mediated workplace, the problem of personal Web usage seems to be related to employees' attitudes. Do they feel that this is an ethical issue? What kind of ethical views do they have on this issue? Above all, how do employees perceive the difference between non-personal Web use and personal Web use of the Internet? In this chapter, we investigate why employees engage in this seemingly unethical behavior and why current organizational efforts are not really effective. We are especially interested in perceptual difference between personal Web usage groups and non-personal Web usage groups in the context of non-work-related usage of the Internet. Data from a field survey of 546 business professionals was analyzed to investigate what causes people, who are not currently engaged in personal Web usage, to perceive intention to do so, and what causes people, who are currently engaged in personal Web usage, to continue their activities.

Drawing on previous studies in behavioral intention and human attitude, literature on general deterrence theory and social learning theory, and literature on moral and ethical decision-making, we develop a comprehensive model to explain personal Web usage and identify effective ways to reduce it. We found that most employees do not feel that personal Web usage is ethically wrong, suggesting that organizations need to define an ethical boundary for Internet use. We empirically confirm that current regulatory efforts, such as rules, policies, or monitoring systems, are not effective in reducing either employee intention to commit personal Web usage or the frequency of the usage, indicating that organizations need to redefine their current counter-measures. Our results also show that employees' current personal Web usage frequency varies depending on their organizational level and years of work experience.

This chapter is organized as follows. First, previous studies of computer abuse are reviewed. Second, research model and hypotheses are discussed. Third, method and the results of our data analysis are addressed. Finally, implications and conclusions are presented.

COMPUTER-RELATED UNETHICAL BEHAVIOR

Computer abuse, which is defined as "any intentional act associated in any way with computers where a victim suffered, or could have suffered, a loss" (Parker, 1998, p. 333), has received much attention from both industry and academia as computers become important resources in organizations. Inundated with problems related to computer abuse, researchers have tried to understand this phenomenon from many different perspectives, especially in the fields of ethics, criminology, and social psychology.

The field of ethics defines an unethical decision as being "illegal or morally unacceptable to the larger community" (Jones, 1991, p. 367), and many theoretical and empirical studies have been done in relation to ethical decision-making at the individual and organizational levels (Bommer et al., 1987; Flannery & May, 2000). These studies dealt with developing an instrument for measuring moral development, building a model for an ethical decision-making process, and finding the factors that affect the unethical behavior such as moral norms (or obligations), moral intensity, and denial of responsibility.

In the criminology field, previous research in computer-related crimes mainly focused on computer abuse issues in relation to general deterrence theory (e.g., Hoffer & Straub, 1989; Parker, 1998; Straub & Nancy, 1990) and social learning theory (Agnew, 1995, 1998; Akers et al., 1979). The studies based on general deterrence theory assume that deviant behavior could be deterred if deviants felt insecure about being detected and punished severely (Straub & Welke, 1998), and recommend deterrence mechanisms such as computer use policies, security systems, and security awareness programs. Social learning theory asserts that a person is involved in computer-related crimes because he or she becomes more likely to associate with delinquent peers who transmit delinquent values, reinforce delinquency, and function as delinquent role models.

In the social psychology field, the theory of reasoned action (TRA) (Ajzen & Fishbein, 1980) and its extended theory of planned behavior (TPB) (Ajzen,

1985) have been applied to a number of ethical decision-making situations to ascertain the relationship between attitude, subjective norms, perceived behavior control, behavioral intention, and behavior. Examples of applying these theories range from business-related ethical issues, such as waste water treatment (Flannery & May, 2000) and information disclosure of financial products (Kurland, 1995), to general cheating behaviors, such as shoplifting (Beck & Ajzen, 1991). More recently those theories have been used for ethical and criminal issues related to computer use (e.g., Banerjee, Cronan, & Jones, 1998).

The theories mentioned above, as well as the empirical investigation of these theories, have provided us with a good understanding of why people are involved in computer abuse and what mechanisms are effective in preventing it. However, we need a more comprehensive picture of end-users' unethical behavior in order to understand personal Web usage since:

1. There have been many studies about unethical (or abusive) behaviors across several fields such as ethics, criminology, and psychology, but they have mainly focused on their independent theoretical points of view to address this issue. Since there were no previous efforts to integrate those different points of view, developing a comprehensive model is a valuable effort to understand personal Web usage of many different types of end-users.
2. While Internet-related unethical (or abusive) behaviors are widespread and exponentially increasing in our daily work, few studies have been performed and empirically validated along with situational factors to address issues related to end-users' personal Web usage. Many researchers indicate that further studies to explain situational unethical behavior in the information technology context are needed.
3. Most studies in ethical decision-making have ignored moral dimensions and simply applied the attitude-intention scheme provided by TRA or TPB. We believe that the inclusion of various factors like moral obligation and denial of responsibility will provide valuable insight in understanding Internet-related moral decision-making problems.

RESEARCH MODEL AND HYPOTHESES

Figure 1 shows our research model. We developed a comprehensive model based on theories from several disciplinary areas. Different types of

Figure 1. Research Model

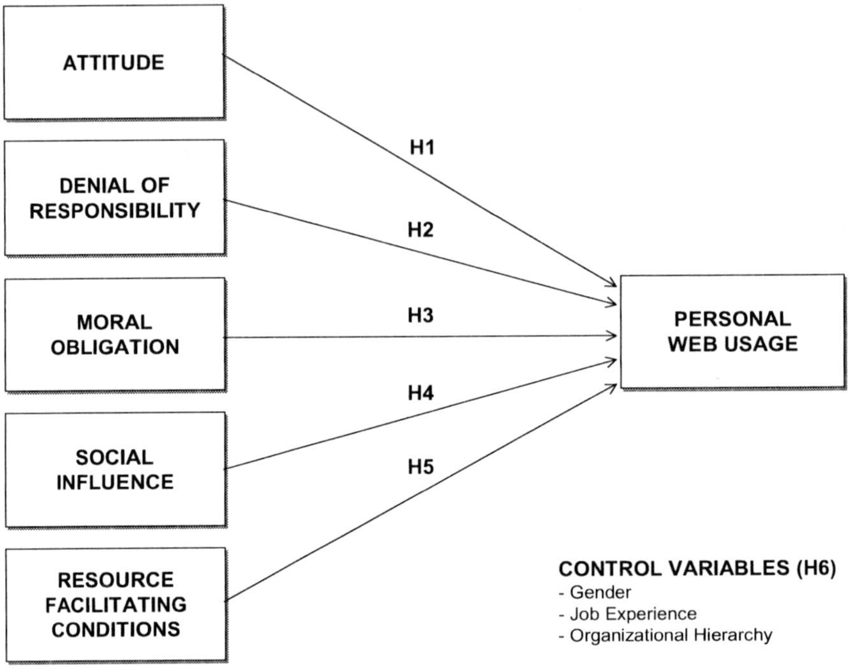

moral attitudes toward personal Web usage were used along with facilitating/deterring situational factors, social influences, denial of responsibility, and moral obligations.

Personal Web Usage

As statistics have indicated that personal use of the Internet at work is a pervasive trend, we raise a question of when it becomes personal Web usage. In our study, we define personal Web usage as "extensive personal use of the Internet at work" on the grounds that most individuals use the Internet for personal purpose and that whether it becomes unethical (or abusive) behavior is a matter of frequency and time spent. For our analysis purpose, we consider non-work-related Internet use for more than 30 minutes a day as extensive personal use since companies are adopting an Internet policy that any extra usage over 30 minutes should be approved by supervisors (Siau, Nah, & Teng, 2002).

Attitude

An attitude toward a behavior is defined as "the degree to which the person has a favorable or unfavorable evaluation of the behavior in question" (Beck & Ajzen, 1991, p. 286). Previous studies have found that attitude significantly affects behaviors. In the ethical context, it has also been known to have significant effect on ethical decision-making (e.g., Flannery & May, 2000). Assuming that personal Web usage is a kind of unethical behavior, we expect that employees' favorable attitude toward personal Web usage will be positively related to personal Web usage (H1).

Denial of Responsibility

The denial of responsibility (RD), defined as people's tendency to ascribe responsibility to oneself or to diffuse and depersonalize it to others, is related to rationalizing the consequences of one's behavior (Harrington, 1996, p. 261). It is known that the lower people's RD is, the more they accept responsibility and feel responsible for others' welfare, while the higher people's RD is, the greater their tendency to ignore social or organizational norms, and to rationalize their unethical behavior by disregarding others like organizations or teammates (Harrington, 1996). In accordance with this prediction, we expect that employees' high RD toward personal Web usage will be positively related to personal Web usage (H2).

Moral Obligation

Recently a number of studies in ethics and moral decision-making have shed light on moral obligation (e.g., Gorsuch & Ortberg, 1983). Moral obligation (or norm) is defined as an individual's perception of the moral correctness or incorrectness of performing a behavior (Corner & Armitage, 1998). Previous studies found that moral obligation was a significant predictor of diversified behaviors. For instance, Kurland (1995) indicated that moral obligation affects insurance agents' ethical intentions. Similarly, Randall and Gibson (1991) provided evidence of moral obligation's significant effect on the behavioral intention of nurses. In the same context, we expect that employees' high moral obligation will be negatively related to personal Web usage (H3).

Social Influence

Social influence is defined as the social pressure to perform or not to perform the behavior. Social influence was found to be an important factor in explaining human intention in the social psychology field, information technology adoption, and computer-mediated communications. Social learning theory in criminology, developed by Akers and his associates (Agnews, 1995; Akers et al., 1979; Akers, 1998), shows how people's inclination to follow significant others (i.e., coworkers and seniors) is related to unethical or criminal behavior. The theory pointed out that there is a highly positive relationship between criminal friends and delinquency, and it has been empirically validated (e.g., Agnew, 1995). In the same context, we expect that employees' perception of significant others' behavior and attitude will be positively related to personal Web usage (H4).

Resource Facilitating Conditions

Whether resources are easily available or not is considered an important factor that governs an individual's behavior (Ajzen, 1985; Ajzen & Fishbein, 1980). Ajzen (1985) argued that judgment of resource accessibility and opportunity for completing unethical behavior successfully, as well as the perceived power of each facilitator or inhibitor of the behavior, would direct human intention. Previous research proved significant roles of resource facilitating conditions in behavioral intention and behavior under various situations (Kimieck, 1992).

General deterrence theory in criminology provides a theoretical ground to address the importance of resource facilitating conditions to unethical behavioral decision. The theory asserts that individuals make rational decisions in order to maximize their benefits and minimize the costs. Therefore, a person can make a criminal decision when the expected benefits caused by the criminal action exceed the cost of punishment. The theory, especially, focuses on the cost factors of deterring criminal behavior through means such as policies, systems, and awareness programs, and it is found that they significantly reduce the criminal intention or behavior (e.g., Loch & Conger, 1996; Straub & Welke, 1998).

In this study, we found five resource facilitating factors that are specific to personal Web usage through extensive interviews with employees in organizations. They are: ease of accessibility of personal computer (PC), seclusion of office, amount of workload, Web usage policy, and network monitoring/ filtering systems. We expect that ease of accessibility of PC (H5A) and seclusion of office (H5B) are positively related to personal Web usage, and amount of workload (H5C), Web usage policy (H5D), and network monitoring/filtering systems (H5E) are negatively related to personal Web usage.

Control Variables

The cause of individuals' different ethical decisions has recently become a popular area of study in the field of ethics. Loe, Ferrell, and Mansfield (2000) performed meta-analysis of individual differences related to ethical decision-making and found that individual characteristics such as gender, age, education, and work experience affect ethical decisions. Similar studies have found that gender, age, and year of experience are significant factors that are related to ethical decision-making (Serwinek, 1992). Consistent with current findings, our model includes gender, year of experience, and organizational hierarchy as control variables. We expect that there is significant gender difference in personal Web usage (H6A), and year of experience (H6B) and organizational hierarchical level (H6C) will be negatively related to personal Web usage.

RESEARCH METHOD

All measurements were developed based on previous studies. We had eight groups of factors: (1) personal Web usage (dependent variable); (2) intention to commit personal Web usage (dependent variable); (3) attitudinal tendency toward personal Web usage; (4) denial of responsibility; (5) moral obligation; (6) social influence; (7) work environment factors — internal control, resource facilitating conditions, and constraining conditions; and (8) control variables.

We used both frequency and the amount of time engaged in non-work related Internet use to measure personal Web usage. We used one item to measure the employee's intention to commit personal Web usage. It was measured using a seven-point Likert scale ranging from 1 (strongly disagree) to 7 (strongly agree). Attitudinal tendency was adapted based on Ajzen (1985). Our measurement for moral obligation was adapted from Tessler and Schwartz (1972) and Banerjee, Cronan, and Jones (1998). Items for social influence were also developed based on previous studies (Ajzen & Fishbein,

1980). We found that coworkers and seniors (or supervisors) were important referents during the testing period. We identified five resource facilitating/constraining factors related to personal Web usage through extensive interviews with 12 employees, and each of them was measured using a single item. The use of situation-specific factors has been advocated to provide a better understanding of interested behaviors (Ajzen, 1985). Finally we used gender, year of job experience, and organizational hierarchy as control variables. The measurement was modified through pre-test and pilot test.

Seven hundred and forty (740) questionnaires were distributed to U.S. business professionals in the northeast coast, and 561 were returned. We removed 15 incomplete and invalid questionnaires, leaving 546 responses for analysis. The overall response rate was 74%.

RESULTS

Initial Statistics

Male respondents represent slightly over half of the sample (50.9%). Sixty-eight percent of respondents are younger than 41. Respondents are evenly distributed in the organizational hierarchy. The average work experience in their line of work is 10.1 years. Sixty-one percent of respondents report that their companies have installed systems to monitor or filter personal Web usage. When asked how often they engage in such activities, 29.5% answer never, and 70.5% answer that they do at least once a month. Since we believe that the factors affect the behavior of employees who currently engage in personal Web usage and those who do not are different, we divided the respondents into two groups — the non-personal Web usage group and the personal Web usage group, based on the 30-minutes-a-day criteria explained in the earlier section. As dependent variables, personal Web usage intention was used for the former, and frequency and amount of time spent for the latter. The intention to commit personal Web usage can be regarded as a strong predictor of future personal Web usage.

Reliability of factors for social influences, attitude, denial of responsibility, and moral obligations were assessed based on Cronbach's α values. All of them were over 0.6, demonstrating reasonable reliability of measures. An inspection of the pair-wise correlation matrix revealed that all correlations among factors used were less than 0.6, confirming the discriminant validity of factors.

Table 1. Regression Results: Non-Personal Web Usage Group vs. Personal Web Usage Group

		Non-personal web usage Group (Intention)		Personal web usage group (Frequency)		Personal web usage group (Time-spent)	
	VARIABLES	β	t-value	β	t-value	β	t-value
FIRST MODEL	GENDER	-.036	-.395	-.026	-.50	.02	.36
	YEAR OF EXPERIENCE	-.104	-1.16	-.212	-3.94**	-.22	-4.0**
	ORG. HIERARCHY	-.060	-.658	.126	2.36**	.04	.69
	R-SQUARE	0.01		0.07		0.05	
TOTAL MODEL	GENDER	.031	.380	-.016	-.312	.005	.1
	YEAR OF EXPERIENCE	-.013	-.165	-.188	-3.60**	-2.21	-4.05**
	ORG. HIERARCHY	-.173	-1.99**	.095	1.83*	.044	.84
	ATTITUDE	.272	2.69**	.021	.338	.058	.91
	DENIAL OF RESPONSIBILITY	.142	1.46	.167	2.85**	.16	2.72**
	SOCIAL INFLUENCE	.097	1.11	.109	1.94*	.10	1.75*
	MORAL OBLIGATION	-.021	-.223	-.058	-.938	.044	.708
	SECLUSION OF OFFICE	-.165	-2.02**	.035	.679	.059	1.12
	AMOUNT OF WORKLOAD	-.226	-2.65**	-.192	-3.73**	-.144	-2.78**
	AVAILABILITY OF PC	-.13	-1.60	-.062	-1.21	-.224	-4.32**
	WEB USAGE POLICY	.102	1.14	-.024	-.464	.024	.456
	MONITORING SYSTEMS	-.077	-.937	-.004	-.082	.120	2.26**
	R-SQUARE	.35		.19		.18	
	R-SQUARE CHANGE	0.34		0.12		0.13	
	F-CHANGE	$6.89_{(9,120)}$ **		$5.48_{(9,330)}$ **		$5.64_{(9,328)}$ **	

*** *Significant at* $\alpha = 0.01$, ** *Significant at* $\alpha = 0.05$, * *Significant at* $\alpha = 0.1$

Multiple ordinary least squares hierarchical regression analysis was used since our variables mixed many categorical and continuous variables. To ascertain the effect of the main variables specified in our hypotheses, all control variables such as gender, year of experience, and organizational hierarchy were first used to explain the variance, and then the main variables were included in the second model to see how much of the variance was further explained. The power of added factors (in the second model) was 85% for the non-personal Web usage group and over 95% for the personal Web usage group, with α = 0.05 and assuming the medium effect. Variance inflation factor (VIF) was used to confirm possible multicollinearity problems, and the index indicated that multicollinearity is not a problem as VIF was less than 10.

FINDINGS

Table 1 shows the regression results. The results are explained separately for the non-personal Web usage and personal Web usage group. For the non-personal Web usage group, the total model explains 34.9% of total variance in the personal Web usage intention, as opposed to only 1.0% explained by the first model based on control variables, with the difference being significant at the 99% level ($F_{(9,120)} = 6.9$). Among all major variables, only a favorable attitude towards personal Web usage (H1) was positively related to respondents' stronger intention, indicating that employees are not significantly affected by denial of responsibility (H2), moral obligation (H3), and significant others' influence (H4) for future personal Web usage.

We also found that ease of accessibility (H5A) does not have a significant effect on the intention. One possible reason for this insignificance is that employees do not feel the difference in the computing environment between home and the workplace with the fast diffusion of personal computers and high-speed networks at home at reasonable costs. Seclusion of workspace (H5B) and workload (H5C) are negatively related to personal Web usage intention, indicating that high workload and lack of seclusion of one's own desk is a serious barrier to the personal Web usage intention. Counter-measures such as personal Web usage policy (H5D) or installation of monitoring systems (H5E) do not affect personal Web usage intention. The results reflect that employees will not consider organizational preventive efforts as dangerous. There was no significant relationship between gender (H6A) and year of experience (H6B) and personal Web usage intention. However, organizational hierarchy (H6C) has a negative relationship with personal Web usage intention, suggesting that the higher in the organizational hierarchical level employees are, the stronger intention they have to commit personal Web usage.

For the personal Web usage group, the first model with the individual variables explains 7.2% of the variance, and the total model explains 19.3% in personal Web usage, and again the difference is significant ($F_{(9,330)} = 5.5$) at the 99% significant level. "Frequency of" and "time spent on" personal Web usage are used as dependent variables for the personal Web usage group. Attitude toward personal Web usage (H1) and moral obligation (H3) do not have significant effect on either frequency or time spent. However, denial of responsibility (H2) and social influence (H4) have significant effect on both frequency and time spent. For resource facilitating conditions, only workload (H5C) is a significant constraining factor on both frequency of and time spent on personal Web usage. Seclusion of office (H5B) and policy (H5D) are not significant factors on both criteria. Meanwhile, ease of accessibility (H5A) and monitoring systems (H5E) showed mixed results. The former was only significant in time spent, and the latter was in frequency. The significant effect of ease of accessibility on time spent indirectly explains the addictive nature of personal

Web usage, meaning that once an employee starts personal Web use, he/she cannot stop doing it for several hours.

Individual information related to organizational variables such as organizational hierarchical level (H6C) and years of experience (H6B) are highly related to personal Web usage frequency, but only organizational hierarchy has a significant relationship with time spent. It indicates that young employees in lower levels in the organizational hierarchy tend to more frequently use the Internet for personal purpose. Gender (H6A) had no significant effect over the non-personal Web usage or personal Web usage group, asserting that like many computer-related behaviors, personal Web usage is gender-neutral.

DISCUSSIONS

Our data, resulting from a large sample of different functional areas across several organizations, provide interesting results and implications in understanding personal Web usage.

First, we find different patterns of attitude and social influence between the non-personal Web usage group and the personal Web usage group. The non-personal Web usage group's behavior is governed by its members' favorable attitude toward personal Web usage, but once they become involved with personal Web usage, frequency and the amount of time spent are determined by their perception of denial of responsibility. Note that the concept of denial of responsibility used in our study is not just a passive inclination toward personal Web usage, but employees' belief that what they are doing in fact contributes to the benefit of the organization. Also we find that social influence is only a strengthening factor, not an initiating factor, in personal Web usage. Although this finding is not related to intention, it is related to both frequency and time spent, suggesting that once people begin to engage in personal Web usage, their frequency and time spent are affected by the people around them. The fact that others are involved in personal Web usage does not give rise to one's intention to do so. These results imply that separate strategies should be applied to the non-personal Web usage group and the personal Web usage group, respectively.

Second, the results of our analysis indicate that employees do not seriously think of personal Web usage as harmful or unethical behavior. Moral obligation is not a factor that constrains Internet use for both non-personal Web usage and personal Web usage groups. This result implies that the behavior related to personal Web usage is determined by employees' perception of how relevant the behavior is, not by the perception of how unethical personal Web usage is. In other words, the reason that non-personal Web users do not commit personal Web usage is not because they feel that personal Web usage is unethical or harmful, but simply because their situations do not allow it. The two most significant deterring factors are that they do not have enough time for personal Web usage (i.e., high workload) and that they do not have a private office space (i.e., seclusion of workspace). This result is consistent with one of the recent studies from the Wharton Forum on Electronic Commerce that reported a negative relationship between personal use of the Internet at work and workload (Bellman, Lohse, & Johnson, 1999).

Third, current policies, or even the installation of monitoring systems against personal Web usage, do not significantly affect either the intention to commit or the frequency of personal Web usage. This is an interesting result considering that most organizations have enforced policies and installed the systems. Only the amount of time spent on personal Web usage was related to the installation of these systems. Employees do not seem to take current Internet policies seriously when they use the Internet for personal purposes.

Fourth, as expected, employees who are relatively lower in the organizational hierarchy and have less work experience tend to engage in personal Web usage more frequently and for longer periods of time. On the other hand, employees who are higher in the organizational hierarchy showed a stronger behavioral intention for personal Web usage, but actually committed less, as measured in duration and frequency. We are not sure why their intention is not related to actual behavior. It might be related to their computer skills or to the nature of their work that does not require extended use of the Internet.

Finally, individual differences in years-of-work-experience and organizational hierarchical level are identified as having partially significant effects. Organizational hierarchy is significant for the non-personal Web usage group, and both year of job experience and organizational hierarchy are significant for the personal Web usage group. Interestingly, organizational hierarchy shows a different effect. The result implies that higher-level employees who do not perform personal Web usage have a greater intention to do so in the future, while lower-level employees tend to be engaged in personal Web usage more frequently and spend more time on it. Gender is not a significant factor both on intention or behavior.

IMPLICATIONS

While business ethics has been a major research area in the management field in order to deal with organizational ethical issues, there have been few studies in the IS field that targeted ethical issues related to information technology use. In light of this situation, this study provides useful insights both theoretically and practically.

First, this study attempts to integrate several theories from the fields of ethics, computer security, morality, and criminology to address the personal Web usage problem. Despite the importance of an understanding of computer and Internet-related unethical (or abusive) behavior, there are few theoretical studies in this area. Some recent studies explain specific IT-related ethical issues using an attitude-behavioral theoretical view (e.g., theory of planned behavior), but we still do not have much knowledge about diverse IT-related ethical behavior. Although there might be a higher risk of increased model complexity, we have tried to integrate as many different factors from various theories in order to acquire a more thorough understanding of the behavior.

Second, we tested factors from ethical decision-making and investigated how moral judgments play a role in personal Web usage. The result showed that moral obligation is not a significant factor in personal Web usage. This is a rather surprising result, considering that it has been reported to be a predictor of moral decision intention. The results from this study, however, are consistent with Flannery and May (2000), who found that moral obligation is of little importance in managers' environmental ethical decision intention in the U.S. metal-finishing industry. We can better interpret our findings from the significant result of denial of responsibility in Internet use behavior. Not only is denial of responsibility a significant predictor of personal Web usage, but also most of the personal Web usage group agreed to the statements "sometimes they find useful information" and "it helps me reduce my work stress and increase work productivity." The results are also closely related to Gattiker and Kelley's (1999) prediction that computer users may not be able to recognize an ethical dilemma given that the computer environment makes it difficult to recognize the material and psychological consequences to other users. The role of personal moral obligation on computer abuse in the computerized environment warrants more research.

Third, the finding that the availability of Internet policies or the installation of monitoring systems is not an effective way to prevent employees from engaging in personal Web usage provides another interesting implication in general deterrence theory and ethical behavior. Previous studies have advocated the use of clear policies and counter-measure systems against computer abuse (e.g., Straub & Welke, 1998). The results of this study, however, indicate that those efforts are not effective. In interpreting these inconsistent findings, we suggest that our results might indicate the inappropriate operation of policies and systems, not the inherent uselessness of current counter-measures; ineffectiveness of regulatory measures can also be explained in the context of unawareness of the contents of the policy, relatively little punishment, the operation of systems without the consent of employees, and the non-existence of or the irregular implementation of awareness programs.

Above all, we expect that all these symptoms are likely to be the result of "the absence of social consensus," where ethical confusion of employees is created by the difference "between the rapid speed of technological advances and the slower speed by which ethical guidelines for utilization of new technologies [are developed]" (Morris & McDonald, 1995, p. 81). The result clearly suggests that organizations should not only set up policies and install monitoring systems, but also diffuse a social consensus concerning the ethics of personal Web usage. These ethical guidelines concerning Internet use should be set up before investing in various counter-measure systems.

Finally, another interesting implication for computer-related ethical decision-making issues is that situational, rather than dispositional, characteristics influence personal Web usage. This study shows that organizational (organizational hierarchy, years-of-work-experience), individual attitude toward personal Web usage (attitude, denial of responsibility), and situational (workload, seclusion of workspace) factors influence personal Web usage, although gender is not a factor. The results lead us to believe that in the domain of personal use of the Internet, where the situation is characterized by people's ambiguity concerning their ethical dilemma, situational factors, rather than demographics or personal characteristics, play a more important role.

CONCLUSION

While companies focus their efforts on reducing personal Web usage by using several counter-measures, it still remains a significant problem. This study found that major problems come from ethical attitude, significant referents of employees, and control factors affecting their behavior. Therefore, companies need to place more emphasis on getting a better understanding of employees' unethical behavior and the specific situational factors that drive their unethical behavior. To influence employees' ethical judgment on Internet use, companies, especially human resource management departments, need to develop detailed ethical guidelines, run special programs for Internet-addicted employees, and maintain an ethical climate for non-personal Web users with formal and informal meetings with their referents in an organization.

In addition, organizations should re-examine and revise their countermeasures with the participation of employees, perform regular awareness programs, and maintain top management support for personal Web usage issues. This study will provide a strong theoretical alternative for future study concerning the ethical issues related to information technology in an organization, and suggests useful guidelines for practitioners who wish to successfully reduce personal Web usage problems using highly limited resources.

REFERENCES

Agnew, R. (1995). Testing the leading crime theories: An alternative strategy focusing on motivational processes. *Journal of Research in Crime and Delinquency*, 32(4), 363-398.

Ajzen, I. (1985). From intentions to actions: A theory of planned behavior. In Kuhl, J. & Beckmann, J. (Eds.), *Action Control: From Cognition to Behavior* (pp. 11-39). Berlin: Springer-Verlag.

Ajzen, I. & Fishbein, M. (1980). *Understanding Attitudes and Predicting Social Behavior*. Englewood Cliffs, NJ: Prentice-Hall.

Akers, R.L. (1998). *Social Learning and Social Structure: A General Theory of Crime and Deviance*. Boston, MA: Northeastern University Press.

Akers, R.L., Krohn, M.D., Lanza-Kaduce, L., & Radosevich, M. (1979). Social learning and deviant behavior: A specific test of a general theory. *American Sociological Review*, 44, 636-655.

Banerjee, D., Cronan, T.P., & Jones, T.W. (1998). Modeling IT ethics: A study of situational ethics. *MIS Quarterly,* 22(1), 31-60.

Beck, L. & Ajzen, I. (1991). Predicting dishonest actions using the theory of planned behavior. *Journal of Research in Personality,* 25, 285-301.

Bellman, S., Lohse, G.L., & Johnson, E.J. (1999). Predictors of online buying behavior. *Communications of the ACM,* 42(12), 32-38.

Bommer, M., Gratto, C., Gravander, J., & Tuttle, M. (1987). A behavioral model of ethical and unethical decision making. *Journal of Business Ethics,* 6, 265-280.

Conlin, M. (2000). Workers, surf at your own risk. *Business Week*, (June 12), 105-106.

Conner, M. & Armitage, C.J. (1998). Extending the theory of planned behavior: A review and avenues for future research. *Journal of Applied Social Psychology,* 28(15), 1429-1464.

Flannery, B.L. & May, D.R. (2000). Environmental ethical decision making in the U.S. metal-finishing industry. *Academy of Management Journal,* 43(4), 642-662.

Ford, R.C. & Richardson, W.D. (1994). Ethical decision making: A review of the empirical literature. *Journal of Business Ethics,* 13, 205-221.

Gattiker, U.E. & Kelley, H. (1999). Morality and computers: Attitudes and differences in moral judgments. *Information Systems Research,* 10(3), 233-254.

Gorsuch, R.L. & Ortberg, J. (1983). Moral obligation and attitudes: Their relation to behavioral intentions. *Journal of Personality and Social Psychology,* 44, 1025-1028.

Hancock, B. (1999). World Wide Web abusive use widespread. *Computers & Security,* 18(3), 195-196.

Harrington, S.J. (1996). The effect of codes of ethics and personal denial of responsibility on computer abuse judgments and intentions. *MIS Quarterly,* 20(3), 257-278.

Hoffer, J.A. & Straub, D.W. (1989). The 9 to 5 underground: Are you policing computer abuses? *Sloan Management Review,* 30(4), 35-44.

Infoworld. (2000, April 19). Employee study cites rampant personal Web usage. Available online at: http://www.infoworld.com/articles/en/xml/00/04/19/000419ensurfers.xml.

Jones, T.M. (1991). Ethical decision making by individuals in organizations: An issue-contingent model. *Academy of Management Review,* 16(2), 366-395.

Kimieck, J. (1992). Predicting vigorous physical activity of corporate employees: Comparing the theories of reasoned action and planned behavior. *Journal of Sport and Exercise Psychology,* 14(2), 192-206.

Kurland, N.B. (1995). Ethical intentions and the theories of reasoned action and planned behavior. *Journal of Applied Social Psychology,* 25(4), 297-313.

Loch, K.D. & Conger, S. (1996). Evaluating ethical decision making and computer use. *Communications of the ACM,* 39(7), 74-83.

Loe, T.W., Ferrell, L., & Mansfield, P. (2000). A review of empirical studies assessing ethical decision making in business. *Journal of Business Ethics,* 25, 185-204.

Morris, S.A. & McDonald, R.A. (1995). The role of moral intensity in moral judgments: An empirical investigation. *Journal of Business Ethics,* 14, 715-726.

Parker, D.B. (1998). *Fighting Computer Abuse — A New Framework for Protecting Information.* New York: John Wiley & Sons.

Randall, D.M. & Gibson, A.M. (1991). Ethical decision making in the medical profession: An application of the theory of planned behavior. *Journal of Business Ethics,* 10, 111-122.

Serwinek, P.J. (1992). Demographic and related differences in ethical views among small business. *Journal of Business Ethics,* 11, 555-566.

Siau, K., Nah, F., & Teng, L. (2002). Acceptable Internet policy. *Communications of the ACM,* 45(1), 75-79.

Straub, D.W. & Nance, W.D (1990). Discovering and disciplining computer abuse in organizations: A field study. *MIS Quarterly,* 14(1), 45-62.

Straub, D.W. & Welke, R.J. (1998). Coping with systems risk: Security planning models for management decision making. *MIS Quarterly,* 22(4), 441-465.

Tessler, R.C. & Schwartz, S.H. (1972). Help-seeking, self-esteem, and achievement motivation: An attributional analysis. *Journal of Personality and Social Psychology,* 21, 318-326.

Chapter III

When Work Morphs into Play: Using Constructive Recreation to Support the Flexible Workplace

Jo Ann Oravec
University of Wisconsin - Whitewater, USA

ABSTRACT

Organizations have become more permeable — integrating more influences from the outside world — as participants engage in such online diversions as trading stocks, engaging in multiplayer games, or viewing images of their children in daycare. Availability of these activities has brought the potential for abuse but also new opportunities. Constructive uses of online recreation and play can enhance many workplaces (especially high-tech and information-saturated ones) and perhaps ultimately make them more productive. Human resource (HR) professionals can become active in exploring and tailoring constructive recreation strategies for specific workplace contexts. Many organizational roles today demand high levels of creativity and mental flexibility, and constructive uses of online recreation can help individuals gain fresh perspectives. This chapter proposes that these complex issues be resolved through participatory approaches, involving workgroups and HR professionals in discussions as to what constitutes "constructive recreation," as well as in development and dissemination of effective and fair organizational policies.

INTRODUCTION

Issues concerning the boundaries between work and play have provided continuing struggles for managers and employees as well as headaches for human resource (HR) professionals. Sociologist Donald Roy (1959-1960) used the "banana time" notion to capture how employees have made workplaces more tolerable by participating in off-task camaraderie. Banana time was the collectively determined break time of factory workers, the start of which was signaled with a lunchbox banana. Industrial economist Robert Schrank (1978) wrote of how "schmoozing" supported the informal organization of workplaces, providing not just recreation but increased levels of workplace cohesion. In the "information age," such playful, exploratory, and spontaneous interaction can also facilitate the exchange of ideas and insights for tackling workplace problems. HR professionals within organizations should have some sense of how online play relates to work (especially knowledge work) so as to increase the productivity and support the well-being of organizational participants.

The Internet has supplied new dimensions to workplace recreation issues. It infuses a bevy of opportunities for diversion into everyday work contexts — although the individuals with whom one "schmoozes" or enjoys "banana time" can be many miles distant. Online games can be seen on workstations in nearly every organization, and growing numbers of employees regularly access online sports scores. Workplaces have become more "porous" and permeable — integrating more influences from the outside world — as individuals engage in such online diversions as trading stocks or viewing images of their children in daycare. Availability of these activities has brought the potential for abuse (as related elsewhere in this book), but also new opportunities. This chapter presents the case that constructive uses of online recreation and play can enhance many workplaces and perhaps ultimately make them more productive.

Everyday workplace life is becoming more diverse and chaotic. Its complex and varying aspects (such as convoluted schedules and malleable timeframes) are often attempts to accommodate massive industrial, technological, and economic shifts (Epstein & Kalleberg, 2001; Gilbert & Bower, 2002; Ofori-Dankwa & Julian, 2001). Although many organizational roles today demand high levels of creativity and mental flexibility, they can also fail to provide the means through which individuals can gain fresh perspectives. Managers who expect employees not to use the Internet for some amount of off-task activity severely misjudge the nature of workplace life — which is solidly infused in online interaction. Depriving employees of opportunities for Internet recreation in some cases excludes the possibility of nearly any form of diversion from assigned responsibilities. This chapter proposes that these complex issues be resolved through participatory approaches, involving workgroups in discussions as to what constitutes "constructive recreation" as well as in development and dissemination of effective and fair policies. This discourse can also ultimately increase levels of trust among team members and between employees and management. Enabling the constructive use of online recreation is certainly not a panacea for workplace ills. However, it can be part of overall strategies to manage people through mutually agreed-upon goal-setting and assessment of outcomes — rather than by what they simply appear to be doing.

SOME BACKGROUND ON THE ISSUES

Workplace use of the Internet for activities that are not directly authorized by management is often considered as the "theft" of human and computer time — comparable to absconding with other forms of organizational resources. Even though many managers consider the personal use of the Internet as an ethical lapse (Greengard, 2000), the "moral high ground" concerning these issues is not entirely clear. Much of the rhetoric and advertising copy associated with workplace computing incorporates recreational imageries and motifs, which can send misleading signals to employees. A number of individuals have already had significant experience combining work with online recreation; convincing them that hard work cannot be combined with online play is thus a tough sell. Telecommuters returning to organizational settings are often not entrusted with the autonomy to engage in online breaks at appropriate times — latitude they take for granted when doing the same tasks in their home offices. Many young people became comfortable with computing through video games and online interpersonal interaction, and took online breaks during their demanding college studies (Colkin & George, 2002). Individuals must find ways to cope psychologically with increased pressures on the job, and management should explore creative but feasible ways to assist them in these efforts.

Wireless Internet applications add more complexities to these issues, further increasing the porousness of organizations and making employees' access to recreation less dependent on systems controlled by their managers. Daniels (2000) reports how wireless technologies (such as PDAs with Internet access) are used even within meetings to amuse and distract participants, often resulting in productivity losses. A number of single- and multiplayer games can be played on cell phones (Schifrin, 2002). Since wireless technologies are still in the early stages of adoption in many organizational contexts, placing severe restrictions on their use (and penalties for misuse) could be counter-productive. Personal computers became familiar workplace additions in the 1980s in part because of their use for gaming, an activity that encouraged employees of a variety of ages and backgrounds to explore the various dimensions of the devices and to become more comfortable with them (Festervand & Meinert, 1994).

If engaged in constructively, online recreation can aid in awakening creativity and increasing well-being, just as appropriate and timely face-to-face diversions have restored employees' energies over the past decades. However, some individuals may not be able to deal with online recreation constructively. They indeed will use it in ways that affect their organizations and themselves negatively, just as some individuals cannot perform adequately on the job for other reasons. Forms of "positive discipline" can be utilized if employees choose to exceed reasonable, agreed-upon limits; implementing such discipline "requires that the supervisor and employee work together to correct the problem behavior" (Guffey & Helms, 2001). Managers and employees should strive together to harness online recreation toward positive ends, rather than condemning or seeking to stifle it completely.

WHAT IS "CONSTRUCTIVE RECREATION"?

Online recreation has already served many supportive purposes in organizations; games can be used to help decrease computer anxiety and encourage experimentation (Agarwal & Karahanna, 2000; Oravec, 1999). What would make online recreation optimally beneficial to individuals, project teams, and the organization as a whole? To start the discussion: recreation is "constructive" when it is in synch with pending work responsibilities, allowing individuals to use time not consumed by workplace demands in ways that equip them to face future tasks with greater energy and expanded perspectives. Constructive recreation is also in keeping with technological constraints, as exemplified by the organizations that allow online recreation but place limits during certain hours to avoid system overload (Gibbs, 1998; Verton, 2000). Policies established are crafted in participatory ways, and are disseminated broadly (such as some of the policies described in Verespej, 2000).

The major impetus behind constructive recreation is in facilitating the rapid adaptation of individuals to changing circumstances. Online recreation and play can provide needed breaks among disparate activities, as well as hone skills that would otherwise be dormant. Constructive recreation affords individuals the means to maintain their flexibility in workplace environments that place increasing demands on their capacities to withstand change. Giddens (1991), Sennett (1997), and others have provided perspectives on how both workplace and home life are being affected by series of rapid changes, often with profound influences on the very structure of individuals' personalities. Individuals without the psychological and social reserves to adapt can suffer damage as they lose a sense of continuity and meaningfulness. Kanter (2002) compares modern organizations with improvisational theatres, requiring chameleon-like adjustments by their participants to sporadic and unpredictable economic alterations. Improvisation is a difficult art even for trained actors and comedians, testing their ability to adapt to unexpected stimuli (Horwitz, 1996).

Change and flexibility are important, but so are some basic cultural values. Workplace recreation is also "constructive" to the extent in which it is responsive to the overall culture of the organization, and sensitive to the needs and values of other organizational participants (including freedom from harassment). Requirements of project team members in terms of scheduling are especially critical to recognize since the synchronization and sustained involvement of everyone are required during critical periods. Along with its other aspects, recreation is constructive if it provides intellectual and psychological stimulation or support, the sustenance often needed to take on tough challenges. "Reclaimed moments" that individuals spend in such activity can allow them to reestablish senses of control in otherwise stressful and constraining contexts. Ability to access such recreation and thus momentarily escape can provide a safety valve for those who face unyielding situations or put in long work hours, thus putting the porousness of today's Internet-supported workplaces to good use.

Many employees work long hours (often voluntarily) and are reluctant to leave their workstations or other network connections for vacations or even for weekends, given increasing levels of competition and economic uncertainty (Deetz, 1995). Knowledge workers often need to accomplish tasks for which strict timeframes are counterproductive (Alvesson, 2000), for example because of time-zone differences among collaborators. An Ipsos-Reid poll relates that approximately 43% of employees claim that they are formally "on call" for extended hours or bring assigned work duties home (Samuelson, 2001). Home life is increasingly hectic as well, and the interaction between work and home life can intensify personal and household stress (Jacobs & Gerson, 2001; Schor, 1991). Workplace absences (especially when they are unscheduled) have a devastating "ripple" effect in organizations (Robinson, 2002), thus affording employees some leeway on-the-job can thus often result in considerable savings of resources.

The value of recreation and play in adult realms is not well-understood. Credible evidence that individuals who engage in online play are more productive or happier than those who do not will probably never be forthcoming — just as research about related workplace issues often tends to be non-conclusive. Play has been given an assortment of definitions in the academic and research literatures (with examinations in the fields of social psychology, philosophy, and anthropology); it is often considered in both its adult and child modes as a "cognitive and symbolic act that is fundamental to the human representational process" (Myers, 1999). Across species as well as cultures, play has been shown to help individuals prepare for the unexpected by presenting varying streams of novel or challenging situations (Spinka, 2001). Play is generally considered as a support for children's intellectual and social development, but its role in adult lives is less clear. Corbell (1999) projects that there are considerable similarities in the kinds of learning that adults and children can gain from gaming, although adults can put these new insights and cognitive patterns to immediate, practical use. For instance, he describes Norwegian decision makers who use simulation gaming for organizational problem solving. Orbanes (2000) describes how the game Monopoly can impart serious business lessons. Research initiatives on what kinds of recreation and play are most efficacious in different workplace environments — as well as on individual and group "play styles" — could enlighten constructive recreation efforts (although they cannot be expected to provide definitive results).

Simulation is indeed an aspect of play that has some direct implications for employee readiness in the workplace, and it has received some research treatment (Myers, 1999). Michael Schrage's (1999) *Serious Play* examines how simulations expand the intellectual capacities of knowledge workers; forms of online play may equip individuals to utilize an organization's "serious" computer simulations more effectively, thus reinforcing skills applicable in many workplace contexts. Many powerful simulation games with societal or political themes are widely available to the public and have considerable audiences; the Sims series and other popular single- and multiplayer games have been used to entertain and educate in a variety of contexts (Moltenbrey, 2002; Pillay, Brownlee, & Wilss, 1999).

FOSTERING SOCIAL CAPITAL THROUGH ONLINE RECREATION

Managers have often used organizationally sanctioned recreation as a perquisite, a bonus for acceptable conduct. It has served as an extension of the workplace, providing new settings for social interaction. One can be cynical about the softball and bowling leagues sponsored by organizations — but they can help provide a form of "social capital," part of the "glue" that holds the at-work community together (Putnam, 2000). Through the past century, many organizations have sponsored picnics and celebrations with the strategy of increasing workplace cohesion.

As employees (including many white collar as well as knowledge workers) telecommute or put in long and irregular hours, the adhesive that binds organizations has been increasingly conveyed through electronic channels. However, it is unclear what kinds of online activity can foster social capital (Uslaner, 2000). Just as human resource experts struggled early in the 20th century to integrate face-to-face recreation into workplace contexts, organizations should attempt similar feats in online realms, thus making online recreation a shared and open resource rather than a secretive endeavor (Oravec, 1996). Unlike many early human relations experiments, the recreational activities involved should be developed in a participatory (rather than patriarchal) fashion. Whether organization-approved fantasy football, discussion group and collaborative filtering forums, joke-of-the-day contests, or other recreations are ultimately successful will depend on how they fit into everyday working experiences.

Constructive use of online recreation can also help to dispel a number of unfortunate and demeaning workplace practices that ultimately serve to erode trust. In many organizational contexts where face-to-face interaction is involved, employees must go through the effort of looking busy when managers are present; they must create an acceptable "work face" that supposedly reflects productive effort. Often, both managers and employees feel that they have to put in extended hours or make other visible sacrifices for the organization, even when these efforts are apparently not needed for organizational productivity (Alvesson, 2000). Arlie Hochschild (1983) provides examples of such forms of "emotional labor." For instance, flight attendants must appear to be welcoming, whatever their current state of emotion; professionals and service personnel in other fields must similarly take on certain sets of facial and behavioral expressions as they present a face to the world (Goffman, 1959). These expressions are considered relevant to job evaluations in many contexts, often in ways not demonstrably related to productivity. Such emotional labor has online correlates: managers who stop workers from playing online games in idle moments and order them to do inessential tasks signal that what is valued is not work itself, but the appearance that people are productively occupied.

Constructing ways of assigning tasks and evaluating employees so that significant and meaningful measures of productivity are involved can lessen this emphasis on the "surface" behavior of employees. The fostering of understandings concerning online recreation can empower individuals to use time constructively (either in productive effort or in recreation) and avoid such demoralizing emotional labor games.

IMPLICATIONS FOR HR PROFESSIONALS: EFFORTS TO CREATE A LEVEL PLAYING FIELD

Human resource professionals often must deal with competing demands to recognize managerial demands for productivity while they consider the personal needs of organizational participants. The "hype" involving computer networking often obscures the complex social issues involved. Even though there are downturns in the high-tech economy, changes in the Internet applications available to employees are still fast paced. By the time research results are available to inform the decision making of HR departments, many of the issues involved will change in character. HR professionals should thus themselves be conversant with Internet applications and be aware of industry trends so as to be ready when new concerns emerge (such as increasingly sophisticated wireless Internet games).

As workplaces have evolved, so have the issues that have divided employers and managers. Some organizations have taken positive steps to help employees deal with workplace and home pressures (Munck, 2001) and have recognized the importance of loyalty (Alvesson, 2000). However, conflict has ensued for decades on an assortment of matters relating to the quality of work life, often leading to dysfunctional confrontations (Edwards, 1978). Today, employees who guess wrong about online recreation standards — or choose to violate them — often pay large penalties, even being demoted or fired. Some managers have devised negative sanctions for these infringements far more severe than those applied to comparable face-to-face interaction. Office workers paging through paper catalogs in idle minutes rarely face the harsh penalties that those caught shopping online often encounter, even though few computer systems can be construed as "overtaxed" by online shopping. For example, Westlake Chemical in West Charles, Louisiana, simply eliminated access to the Internet to hundreds of employees when managers discovered how much unauthorized Internet activity was going on (Sloan & Yablon, 2000). Companies have encountered considerable penalties as well: Microsoft agreed to a $2.2 million settlement in a sexual-harassment suit involving pornographic messages distributed in an organizational e-mail (Verespej, 2000).

Hard-line positions against forms of online recreation may be required in some instances and directly related to important organizational goals. For instance, air traffic controllers should be expected to keep focused on landing real airplanes rather than escape into fantasy games during assigned hours. However, some hard-line restrictions can reflect fear or lack of understanding of online realms. Management may assume that online recreation will foster or encourage Internet addiction or related concerns. "Internet addiction" has become a widely identified syndrome, although its medical underpinnings are still in question (Beard, 2002; Oravec, 2000). The kinds of non-work activities that are allowed in organizations often mirror managerial culture and values, from softball teams to holiday celebrations. Hard-line restrictions against online recreation and the monitoring of workstations to implement them are of symbolic importance, signaling to organizational participants the "proper" way to view the online workplace and themselves as human beings. Overly restricting online recreation may prevent employees from exploring the full potential of the Internet for productive intellectual and social endeavors. However, a laissez-faire approach may also serve to demoralize workplaces by allowing some individuals to exploit the diligence of team members and possibly even disturb the sensibilities of unfortunate onlookers.

Ambiguities concerning online work and play in virtual realms are increasingly adding complexities to these issues (Broadfoot, 2001). It is often difficult to tell which websites are related to business needs and which are recreational; many have dual purposes, combining amusement with news and other serious pursuits. Slashdot.org has humorous material as well as valuable technical commentary, and abcnews.com has stories on upcoming movies as well as current economic results. Helpful intelligent agents (some with cartoon-like manifestations) can add levity to everyday tasks. Surfing the Internet for an answer to a question or fiddling with various programs can interfere with productive effort, as individuals dwell on technological nuances. Perfecting an organizational newsletter's format can be so involving that individuals lose a sense of proportion as to its business relevance. Managers and employees need to deal not only with recreational concerns but also with broader issues of how to integrate computing into workplaces in ways that are engaging yet productive.

Workplace realities have changed in a tightening economy, and few expect that stability and continuity will replace flux. For many employees the social and recreational activities that are needed for them to function optimally have to be obtained during breaks and unoccupied moments in the workplace rather than after-work initiatives. Many employees (especially in high-tech fields) are on call for long periods, with their know-how required for troubleshooting networks or debugging software programs. Online recreation is part of some individuals' efforts to make these lengthy and demanding working hours more tolerable. A number of online recreational activities can be conducted while productive activity is going on, in a kind of human multitasking. Such multitasking can provide problems if individuals overreach their capacities, in ways comparable to the problem of drivers who engage in cell phone conversations on the road (*Consumer Reports,* 2002). Individuals can check online sports scores while on hold for a telephone call, which can relieve frustration. However, online recreation should not be exploited as a means to keep individuals glued to workstations for indefinite periods in lieu of reasonable work schedules and functional work-life balances.

Solutions as to how to couple online work and play are emerging in organizations that are tailored to specific workplace contexts. Managers and employees are gaining important experience in resolving these issues as individuals perform activities away from direct supervision via mobile computing or virtual office configurations. Managers are learning how to perform their functions without direct employee surveillance. Employees are learning higher levels of self-discipline and the skills of balancing online work and play—just as they have learned to balance face-to-face schmoozing with task orientation in the physical world. Thus setting severe restrictions on online recreation can serve to slow down the process of understanding how to migrate the organization into virtual realms and establish trust. Responsibility and respect for others in these realms can be difficult to acquire, and many employees will indeed need direction. Those who stray from "netiquette" standards in online discussions are generally given guidance as to how they have deviated. Similar kinds of community and peer support will help individuals use recreation constructively in online contexts.

CONCLUSION: MANAGING CONTRADICTION AND PARADOX IN A CHANGING WORKPLACE

The importance of recreation and play is widely recognized for children, but is only slowly being understood in adult realms. Pat Kane has proposed that a "play ethic" be fostered that accommodates the adult requirement for play (www.theplayethic.com; see also Abrams, 2000). Perhaps, given the theme of this essay, a "work/play ethic" is more appropriate, fostering a balance between effort that is immediately productive and other forms of human expression. The notion of accommodating both work and play in organizations can seem paradoxical. In this regard, it joins a number of other paradoxes to be found in organizational contexts, including that of facilitating managerial control as well as employee participation (Stohl & Cheney, 2001). Unfortunately, consensus about the role of play in workplaces is still rare, and human resource professionals must be vigilant for emerging problems and controversies. As evidenced by the accounts in this book, Internet recreation provides a contested space in many organizational settings. This space is quickly expanding as wireless Internet access becomes ubiquitous and as computing equipment becomes pervasive in workplaces.

Allowing for reasonable and humane amounts of online recreation can indeed have considerable advantages, both for the individuals involved and the organization as a whole. It can serve to open blocked creative channels and possibly relieve stress as well. Online recreation can also extend the limits of individuals' working days by providing extra dimensions to workplace activity. Rather than going through the emotional labor of looking busy, employees can utilize spare moments on the job in recharging their mental batteries. Constructive use of recreation will require a number of changes, such as increases in managerial flexibility and employee empowerment (as described as the "new employment relationship" outlined in Boswell, Moynihan, Roehling, & Cavanaugh, 2001). Organizational participants must learn how to handle the distractions and opportunities of increasingly porous workplaces, with their many external influences. Education and training provided by HR professionals can be useful in these initiatives: novice employees can be aided to couple work and recreation in ways that increase overall effectiveness. Constructive recreation strategies can bring these complex matters into the open, rather than allow them to be objects of rumor and fear. Rumor in organizations can have the effect of distorting the issues involved (Scheibel, 2000), making knowledge and power imbalances the primary items of contention rather than the issues at hand.

Forms of online diversion are already becoming integral elements of everyday workplace life, often serving to humanize and enhance organizations. Negotiation and discourse on constructive recreation issues can increase mutual trust and respect concerning online as well as face-to-face activity. With effort on everyone's part (and the coordination strategies of human resource professionals), the constructive use of online recreation can help the entire organization work harder and play harder.

REFERENCES

Abrams, R. (2000). Let's all go out to play. *New Statesman,* 129(4512), 36-37.

Agarwal, R. & Karahanna, E. (2000). Time flies when you're having fun: Cognitive absorption and beliefs about information technology usage. *MIS Quarterly,* 24(4), 665-695.

Alvesson, M. (2000). Social identity and the problem of loyalty in knowledge-intensive companies. *Journal of Management Studies,* 37(8), 1101-1125.

Beard, K. (2002). Internet addiction: Current status and implications for employees. *Journal of Employment Counseling,* 39(1), 2-12.

Boswell, W., Moynihan, L., Roehling, M., & Cavanaugh, M. (2001). Responsibilities in the 'New Employment Relationship': An empirical test of an assumed phenomenon. *Journal of Managerial Issues,* 13(3), 307-328.

Broadfoot, K. (2001). When the cat's away, do the mice play? Control/ autonomy in the virtual workplace. *Management Communication Quarterly,* 15(1), 110-115.

Colkin, E. & George, T. (2002). Teens skilled in technology will shape IT's future. *InformationWeek,* 881(March 25), 72-73.

Consumer Reports. (2002). The distraction factor. 67(2), 18-22.

Crockett, R. (2001). Game theory: Play pays. *Business Week,* 3720(February 19), EB12.

Daniels, C. (2000). How to goof off at your next meeting. *Fortune, 142*(10), 289-290.

Deetz, S. (1995). *Transforming Communication, Transforming Business: Building Responsive and Responsible Workplaces.* Cresskill, NJ: Hampton Press.

Edwards, R. (1978). *Contested Terrain: The Transformation of the Workplace in the Twentieth Century.* New York: Heineman.

Epstein, C. & Kalleberg, A. (2001). Time and the sociology of work. *Work & Occupations,* 28(1), 5-17.

Festervand, T. & Meinert, D. (1994). Older adults' attitudes toward and adoption of personal computers and computer-based lifestyle assistance. *Journal of Applied Business Research,* 10(2), 13-23.

Gibbs, M. (1998). Employees at play. *Network World,* (July 6).

Giddens, A. (1991). *Modernity and Self-Identity: Self and Society in the Late Modern Age.* Cambridge: Polity Press.

Gilbert, C. & Bower, J. (2002). Disruptive change: When trying harder is part of the problem. *Harvard Business Review,* 80(5), 95-101.

Goffman, E. (1959) *The Presentation of Self in Everyday Life.* Harmondsworth: Penguin.

Greengard, S. (2000). The high cost of cyberslacking. *Workforce,* 79(12), 22-23.

Guffey, C. & Helms, M. (2001). Effective employee discipline: A case of the Internal Revenue Service. *Public Personnel Management,* 30(1), 111-128.

Hochschild, A. (1983). *The Managed Heart.* Berkeley, CA: University of California Press.

Horwitz, S. (1996). Improving on a good thing: The growing influence of improvisation. *Back Stage,* 37(30), 22-27.

Jacobs, J. & Gerson, K. (2001). Overworked individuals or overworked families? *Work & Occupations,* 28(1), 40-64.

Kanter, R. (2002). Improvisational theater. *MIT Sloan Management Review,* 43(2), 76-82.

Moltenbrey, K. (2002). Stalking the mainstream. *Computer Graphics World,* 25(4), 26-31.

Munck, B. (2001). Changing a culture of face time. *Harvard Business Review,* 79(10), 125-131.

Myers, G. (1999). Simulation, gaming, and the simulative. *Simulation & Gaming,* 30(4), 482-490.

Ofori-Dankwa, J. & Julian, S. (2001). Complexifying organizational theory: Illustrations using time research. *Academy of Management Review,* 26(3), 415-431.

Oravec, J. (1996). *Virtual Individuals, Virtual Groups: Human Dimensions of Groupware and Computer Networking.* New York: Cambridge University Press.

Oravec, J. (1999). Working hard and playing hard: Constructive uses of online recreation. *Journal of General Management*, 24(3), 77-89.

Oravec, J. (2000). Internet and computer technology hazards: Perspectives for family counselling. *British Journal of Guidance and Counselling,* 28(3), 309-324.

Orbanes, P. (2002). Everything I know about business I learned from MONOPOLY. *Harvard Business Review,* 80(3), 51-58.

Pillay, H., Brownlee, J., & Wilss, L. (1999). Cognition and recreational computer games: Implications for educational technology. *Journal of Research on Computing in Education,* 32(1), 203-217.

Putnam, R. (2000). *Bowling Alone: The Collapse and Revival of American Community.* New York: Simon & Schuster.

Robinson, B. (2002). An integrated approach to managing absence supports greater organizational productivity. *Employee Benefits Journal,* 27(2), 7-12.

Roy, D. (1959-1960). "Banana time": Job satisfaction and informal interaction. *Human Organization,* 18, 158-68.

Samuelson, R. (2001). Fun ethic vs. work ethic? *Newsweek,* 138(11), 43.

Scheibel, D. (1999). 'If your roommate dies, you get a 4.0': Reclaiming rumor with Burke and organizational culture. *Western Journal of Communication,* 63(2), 168-193.

Schifrin, M. (2002). Best of the Web. *Forbes ASAP*, (Summer), 65-84.

Schor, J. (1991). *The Overworked American.* New York: Basic Books.

Schrage, M. (1999). *Serious Play.* Cambridge, MA: Harvard Business School.

Schrank, R. (1978). *Ten Thousand Working Days.* Cambridge, MA: MIT Press.

Sennett, R. (1998). *The Corrosion of Character: The Personal Consequences of Work and the New Capitalism.* New York: W.W. Norton.

Sloan, P. & Yablon, M. (2000). New ways to goof off at work. *U.S. News & World Report,* 129(9), 42-43.

Spinka, M. (2001). Mammalian play: Training for the unexpected. *Quarterly Review of Biology,* 76(2), 141-169.

Stohl, C. & Cheney, G. (2001). Participatory processes/paradoxical practices. *Management Communication Quarterly,* 14(3), 349-408.

Uslaner, E. (2000). Social capital and the Net. *Communications of the ACM,* 43(12), 60-64.

Verespej, M. (2000). Inappropriate Internet surfing. *Industry Week/IW,* 249(3), 59-63.

Verton, D. (2000). Employers OK with e-surfing. *Computerworld,* 34(51), 1-2.

Chapter IV

A Multidimensional Scaling Approach to Personal Web Usage in the Workplace

Murugan Anandarajan
Drexel University, USA

Patrick Devine
Drexel University, USA

Claire A. Simmers
Saint Joseph's University, USA

ABSTRACT

In this study, a typology of workplace personal Web usage (PWU) behaviors was developed using multidimensional scaling techniques. Results suggest that personal Web usage behaviors vary along two dimensions: opportunities versus threats and organizational versus interpersonal. On the foundation of these two dimensions, PWU behaviors appear to fall into four distinct categories: disruptive, recreational, personal learning, and ambiguous PWU. This typology should prove useful for developing conceptual and empirical research agendas of PWU behavior in the workplace.

INTRODUCTION

Reports indicate that about 55 million people in the United States access the World Wide Web ("the Web") from their workplace on a daily basis (Horrigan, 2002). A Department of Commerce study indicates that Web usage in the workplace has a growth rate of approximately 54% per year (U.S. Department of Commerce, 2002). While such growth has the potential to increase worker productivity, it is not without significant problems (Lim et al., 2002; Simmers, 2002). The American Management Association indicates that more than 50% of all workplace-related Web activities are personal in nature (Greengard, 2000). A recent study indicates that, on average, employees spend 8.3 hours a week surfing the Web for non-work-related activities (Websense, 2002). These activities include online entertainment, reading news, making travel arrangements, online purchases, and searching for jobs. Such activities translate into billions of dollars a year in revenue lost due to lost productivity (Mills et al., 2001).

In addition to the costs incurred due to losses in productivity, personal Web usage has caused organizations to face a host of other detrimental issues (Siau & Nah, 2002). There is an increased burden on company servers as bandwidth and system storage gets clogged with non-work-related files (Mills et al., 2001). Organizations also face heightened security risks from viruses and other malicious programs inadvertently downloaded by employees as they use the Web for personal reasons (Sloane, 2002). The costs of such security risks are significant, with an estimated worldwide economic impact of approximately $13.2 billion for 2001 (Computer Economics, 2002). In addition to security costs, companies also face innumerable legal costs as a result of issues ranging from copyright infringement to sexual harassment lawsuits (Roberts, 1999; Panko & Beh, 2002). Personal Web usage is increasingly becoming an issue which management cannot ignore (Simmers, 2002).

Organizations have attempted to respond to the challenges of personal Web usage with policies that range from laissez faire to zero tolerance (Urbaczewski & Jessup, 2002); yet, merely establishing a policy does not adequately address this challenge (Anandarajan, 2002). Management must make an effort to understand the dimensions underlying personal Web usage behaviors if they are to hope to effectively manage workplace Web usage (Lim, 2002). Currently very little research has specifically addressed personal Web usage behaviors (Anandarajan, 2002; Lim, 2002). The goal of this research is to assist in the development of a framework that may be used to categorize personal Web usage behaviors and further our understanding of them.

PERSONAL WEB USAGE

Personal Web usage (PWU) can be defined as *"voluntary online Web behaviors during working time using any of the organization's resources for activities outside current customary job/work requirements"* (Simmers & Anandarajan, 2002). Such online Web behaviors include a wide scope of Web-related activities, such as searching for information, playing games, and communicating in chat rooms (see appendix for the complete list of behaviors). The information systems literature on Web usage has shown a disproportionate emphasis on the desirable Web usage benefits (Anandarajan et al., 2000; Lederer et al., 2000; Teo & Lim, 1998) and undesirable side of Web usage behavior (Griffiths, 1998; Joinson, 1998; Putnam et al., 2000; Lim et al., 2002; Lim, 2002). Other studies have looked at the demographic and motivational variables associated with Internet usage in general without focusing on PWU and its underlying factors (Teo et al., 1999). The few studies that do not fall into the latter two categories have dealt with identifying the types of websites accessed during PWU (Anandarajan et al., 2000; Teo et al., 1999), on the time spent on PWU (Armstrong et al., 2000; Korgaokar & Wolin, 2002; Teo et al., 1999). While specific measures such as sites visited and time spent may serve as useful first steps towards exploring PWU, the majority of these measures are one-dimensional in nature. It is one of the goals of this research to explore the multidimensional nature of personal Web usage and explore the individual behaviors that comprise this usage.

A typology of PWU can be a useful starting point for developing a systematic research agenda. Such a typology can be useful for the development of broader measures of PWU, since such aggregated measures are more reliable and valid than specific measures (Rushton et al., 1983). Through the utilization of user interviews, survey methodology, multidimensional scaling, and cluster analysis, the current study hopes to provide such a typology.

METHODS AND RESULTS

Multidimensional scaling (MDS) is a useful tool for producing inductive, but empirically derived typologies. The advantage of using MDS stems from its ability to determine the dimensions along which a set of stimuli is perceived to vary without large numbers of participants, as is required by factor analysis. Additionally, MDS can be used to discern how participants attend to different dimensions in making judgments (O'Hare, 1976). Another advantage of using MDS is that these dimensions are generated by the participants, not the researcher. Hence, MDS-based typologies are less prone to researchers' biases than typologies developed through other methods. The procedures followed for each phase in this study and the results of each phase are discussed below.

Phase 1: Interviews

Five-hundred-and-twelve (512) part-time MBA students from a university in the northeast United States were asked to describe: (1) their perceptions of PWU, and (2) two examples of PWU behaviors while at work. Next, the first author and a research assistant independently removed redundant words and phrases, and rephrased the descriptions the respondents provided to simplify them, and to ensure that the descriptions were relatively generic and applicable across organizations and occupations. A final pool of statements delineating 50 PWU behaviors was obtained.

Phase 2: Pilot Study

Forty-two (42) undergraduate students were given a survey containing the list of 50 PWU behaviors and a brief description of a target behavior, which appeared at the top of the first page. The respondents rated each PWU behavior in terms of its similarity to or difference from the target behavior, using a nine-point Likert-type scale (1 = very similar, 9 = very different). Generally speaking, multidimensional scaling requires having subjects compare and contrast every possible pair of stimuli [n(n - 1)/2]. For the current study this would have involved subjects reviewing 1,225 comparisons, which would then lead to respondent attrition, errors, and fatigue. A valid means of overcoming this potential difficulty was through having subjects make only a subset of comparisons (Thompson, 1983).

In addition, the respondents were also asked to describe their reasoning in comparing the PWU behaviors and the target behavior using a five-point bipolar scale with the following attribute anchors: *serious loss of productivity/not a serious loss of productivity, not serious waste of time/serious waste of time, low relaxation value/high relaxation value, low learning opportunity/high learning opportunity, not harmful to the company/harmful to the company,* and *not harmful to others/harmful to others*. These criteria

Table 1. Demographics of Sample

Gender	*No.*	*%*
Male	51	41.80%
Female	71	58.20%
Age	*No.*	*%*
18-23	8	6.56%
24-29	39	31.97%
30-34	38	31.15%
35-39	21	17.21%
40-44	10	8.20%
45-49	2	1.64%
50-54	4	3.28%
above 55	-	-
Organizational Position	*No.*	*%*
Top-level Managers	7	5.74%
Middle-level Managers	18	14.75%
Lower-level Managers	13	10.66%
Professionals	48	39.34%
Administrative Staff	20	16.39%
Others	16	13.11%

were the most frequently mentioned in the interviews referred to in Phase 1. These behaviors were tested (the statistical procedure is explained in Phase 3) and based on the results and feedback; the list of behaviors was reduced to 39.

Phase 3: Full Study

Sample. There were 122 respondents, 51 men and 71 women, all of whom were part-time evening students in an MBA program at a northeastern university. All the respondents worked full time. Their average age was 32 years; 31% of the participants were managers, while 39% were professionals, and 16% worked in administrative support. Table 1 provides specific facts about this sample.

Procedures. Each respondent was given a survey containing the list of 39 PWU behaviors and a brief description of a target behavior, which appeared at the top of the first page. The respondents rated each behavior in terms of its similarity to or difference from the target behavior, using a nine-point Likert-type scale. The respondents were also asked to specify the criteria they used to distinguish between the target behavior and each of the PWU behaviors.

A multidimensional solution to the similarity data was sought using the ALSCAL procedure in SPSS v10. This procedure derives spatial configurations of behaviors on the basis of the perceived differences between the behaviors. First a similarity matrix was created by computing the perceived differences between the pairs of PWU behaviors descriptions (Kruskal & Wish, 1978). The similarity matrix was scaled in one to four dimensions. The dimensionality of the data was assessed using the stress index to determine which map configuration explained the most variance. This index, which is a goodness of fit measure, indicates how well data fit a particular configuration, i.e., the higher the stress, and the poorer the fit.

Results. A scree test was created by plotting the stress indices for all four configurations (see Schiffman et al., 1981, for further information on stress tests). The one-dimensional solution had a stress index of .494, and the index dropped considerably to .041 for the two-dimensional solution. This suggests that the two-dimensional solution provided the most parsimonious and accurate description of the data. Figure 1 shows the two-dimensional configuration.

Phase 3: Interpreting the Configuration

Procedures: A formal way of interpreting a configuration is to use a regression technique known as Property Fitting (ProFit) to assess relationships between the attributes and the two-dimensional configuration. The mean of each of the attribute-behavior (dependent variable) was regressed with the coordinates of the two dimensions (independent variables). Separate regressions were performed for each attribute. A positive value for the regression coefficient indicates higher ratings of the attributes, while a negative value indicates that high rating of attributes are associated with negative values of the dimension.

Results: Results of the ProFit analysis are summarized in Table 2. This table gives the regression coefficients for each of the attributes. These regression coefficients are components of the directional vector associated with the ProFit line. Thus, a high value for a regression coefficient associated with a particular dimension indicates that the attribute influences interpretation of a

Figure 1. Two-Dimensional Configuration of PWU Behavior with ProFit Analysis

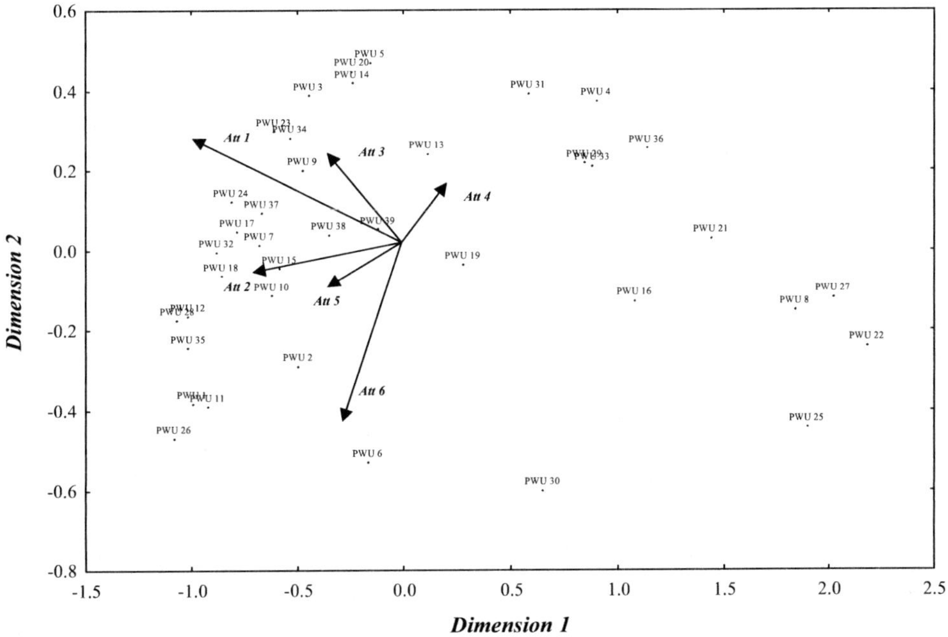

dimension. All F values are significantly different from zero at the 0.0001 level, except for harmful to others, implying that each attribute has a contribution to make to the interpretation of the configuration (Schiffman et al., 1981). This interpretative power tends to be high as indicated by the R^2 values. Four of the attributes of the R^2 range from 0.6 to 0.8. The exceptions are 'relaxing' and 'harmful to others,' indicating these attributes were unlikely to influence the dimensions.

Examination of the beta weights from the regression analysis (Kruskal & Wish, 1978) indicates that there are two most important criteria used by respondents to situate PWU behaviors in the configuration space (Schiffman et al., 1981) and form roughly perpendicular axes.

Dimension 1: Examination of the beta values from the ProFit analysis indicates that *not a serious loss of productivity/serious loss of productivity* (-.88) and *not serious waste of time/serious waste of time* (-.85) and learning (.84) explained the most variance for Dimension 1. Productivity and waste of time were negative, while learning was positive. Since a serious loss

Table 2. Derivation of Labels for the Two Dimensions

Attributes	Dim 1 β_1	Dim 2 β_2	R^2
1 Not a serious loss of productivity/Serious loss of productivity	-0.889 ***	-0.215	0.810
2 Not serious waste of time/serious waste of time	-0.857 ***	-0.137	0.610
3 Low relaxation value/high relaxation value	-0.243	0.224	0.320
4 Low learning opportunity high learning opportunity	0.844 ***	0.222	0.873
5 Not harmful to the company/harmful to the company	-0.235	-0.488 ***	0.781
6 Not harmful to others/harmful to others	-0.223	0.276 *	0.310

of productivity and a serious waste of time are threats to organizations and individuals, we labeled this end of the dimension 'threats.' On the positive side of this dimension, productivity and learning attributes reflected personal usage behaviors that were beneficial, suggesting a label of 'opportunities.' Consequently, we labeled this first dimension "threats versus opportunities of workplace personal Web usage behavior."

Dimension 2: Beta values indicate that *not harmful to the company/harmful to the company* (-.488) and *not harmful/harmful to the individual* (.27) attributes explained the most variance for Dimension 2. One end of this dimension indicated behaviors that impacted the organization, and the other end reflected behaviors that impacted individuals. Personal Web usage behaviors that fell on the negative side were related to organizations and PWU behaviors that were on the positive side were more directly related to individuals. Thus we labeled Dimension 2 "organizational versus interpersonal workplace personal Web usage behaviors."

Phase 4: Locating the Cluster Boundaries and Labeling the Quadrants

Procedures: While MDS and ProFit are useful techniques for mapping out the relationships between the various PWU behaviors, and the nature of the dimensions, these techniques are less useful in identifying the precise location of the boundaries. As suggested by Punj and Steward (1983), a two-stage approach was used to partition the multidimensional scaling map into a number of clusters. First, a hierarchical agglomerative procedure was used to locate the clusters (see Figure 2). The distance between clusters was calculated using Ward's minimum variance method. To validate the cluster solution, a non-

Figure 2. Two-Dimensional Configuration of PWU Behavior with Cluster Analysis

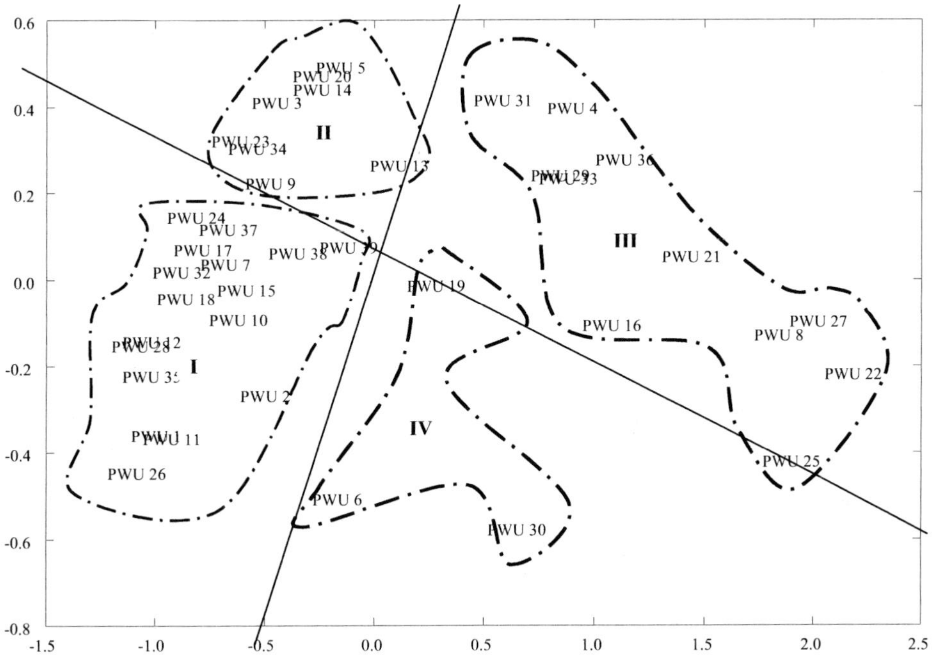

hierarchical analysis was performed, using the K-means method, using the cluster centroids obtained from Ward's method (Hair et al., 2000).

Results: The agglomeration schedule showed that there is a fairly large increase in the value of coefficient from a four-cluster solution to a three-cluster solution, supporting the choice of the four-cluster solution. As with Ward's method, the K-means method had four clusters. This four-cluster solution, which is shown in Figure 3, is very similar to the results produced by the ProFit analysis. It divides the configuration almost exactly along the axes. The four clusters are named and described here:

Cluster 1 — Disruptive PWU

The largest number of behaviors appeared in the lower left quadrant between the "harmful to organization" and "threats" extremes of the axes. These behaviors follow what is commonly viewed as the negative aspects of PWU, sometimes also referred to as Web abuse or cyber-slacking (Siau & Nah, 2002). Behaviors in this grouping include*: visiting adult websites,* *playing online games*, and *downloading music*. These behaviors may have a number of negative consequences for the individual, but the organizations have the greatest exposure to risk with these behaviors. Organizations that do not prohibit many of these behaviors may be accused of de-facto endorsement and held legally, criminally, and financially liable in harassment and discrimination suits (Elron Report, 2000; Mills, 2001). The downloading of software, video, and music has a host of ramifications for the company that range from exposing them to potential computer viruses (Poster, 2002) to clogging up bandwidth and wasting storage space (Mills, 2001; Siau & Nah, 2002). A recent study from SurfControl illustrates this problem:

> *"...in May 2000, nearly 2 million people logged on to a 25-minute Victoria's Secret fashion show Webcast during business hours. The total bandwidth cost for the Webcast was 300,000,000,000 kilobits. If all the viewers watched the show at a modest download rate of 100 kbps, the bandwidth used would have been equivalent to the capacity of nearly 200 million T1 lines."*

Cluster 2 — Recreational PWU

The second cluster is in the top left-hand quadrant. This cluster is bordered by the axes "threats" and "interpersonal." This cluster comprises behaviors such as: *purposeless surfing of the Web*, *searching for weekend recreational/social activities*, *exploring my hobbies/interests*, and *finding information about products I wish to purchase*. This grouping contained leisure and entertainment PWU behaviors, and we label it "recreational usage." We define the usage as engaging in recreational use that is minor, while putting the user's reputation and job security at risk. It seems that the risks of these behaviors far outweigh the benefits derived from them.

Cluster 3 — Personal Learning PWU

The third cluster is in the top right-hand corner and shares the dimensions of interpersonal and opportunities. Behaviors found in this cluster include: *searching for news about my organization, learning about educational/training classes, visiting professional associations' websites,* and *reading about current events*. This cluster may have many indirect benefits for the organization. Seemingly focused more on the acquisition of information and knowledge, and the usage of the Internet as a tool to acquire this information, the cluster may assist in the employee learning process and make them more efficient and effective "surfers" (Oravec, 2002). These behaviors may also serve to be a reduction in stress as employees take a "time-out" from their work-related duties to pursue their own informational inquiries. Thus, both outcomes may lead indirectly to increases in productivity. This cluster may also contain some direct benefits to productivity as well. Many of these behaviors involve searching for and processing information in the form of educational material, current events, news, and company information. These behaviors make for a more informed, better-educated employee, which may actually increase productivity in terms not only of quantity of work performed, but quality as well. Going beyond the scope of this chapter, it is suspected that certain of these benefits are dependent on the type of work as well as the duration of the PWU behaviors the employee is engaged in. Belanger and Van Slyke (2002) note this when they discuss how with certain applications, the learning benefits decline over time as the employees continue to utilize the technology.

Cluster 4 — Ambiguous PWU

The last cluster is the smallest and most ambiguous, and therefore is the most challenging to interpret. This paradoxical grouping contains only three behaviors: *discussing my company in a chat room, looking at government websites,* and *learning about other companies in a chat room.* An example can be seen in the two behaviors that involve chat rooms. These behaviors have the potential to cause harm to the company through statements made by the employees. The company may be held accountable if the employee revealing confidential or damaging information slanders the company or its competitors. However, it is also possible that competitive intelligence might be gained in these "conversations." Though bordered by opportunities and organizational attributes, generalities made from such a small grouping are suspect, and additional work is needed before this cluster can be described with confidence. Figure 3 summarizes the clusters and PWU behaviors.

DISCUSSION

The focus of this study was to contribute to the field of knowledge pertaining to personal Web usage in the workplace. Many studies have been

Figure 3. Typology of Workplace Personal Web Usage Behavior

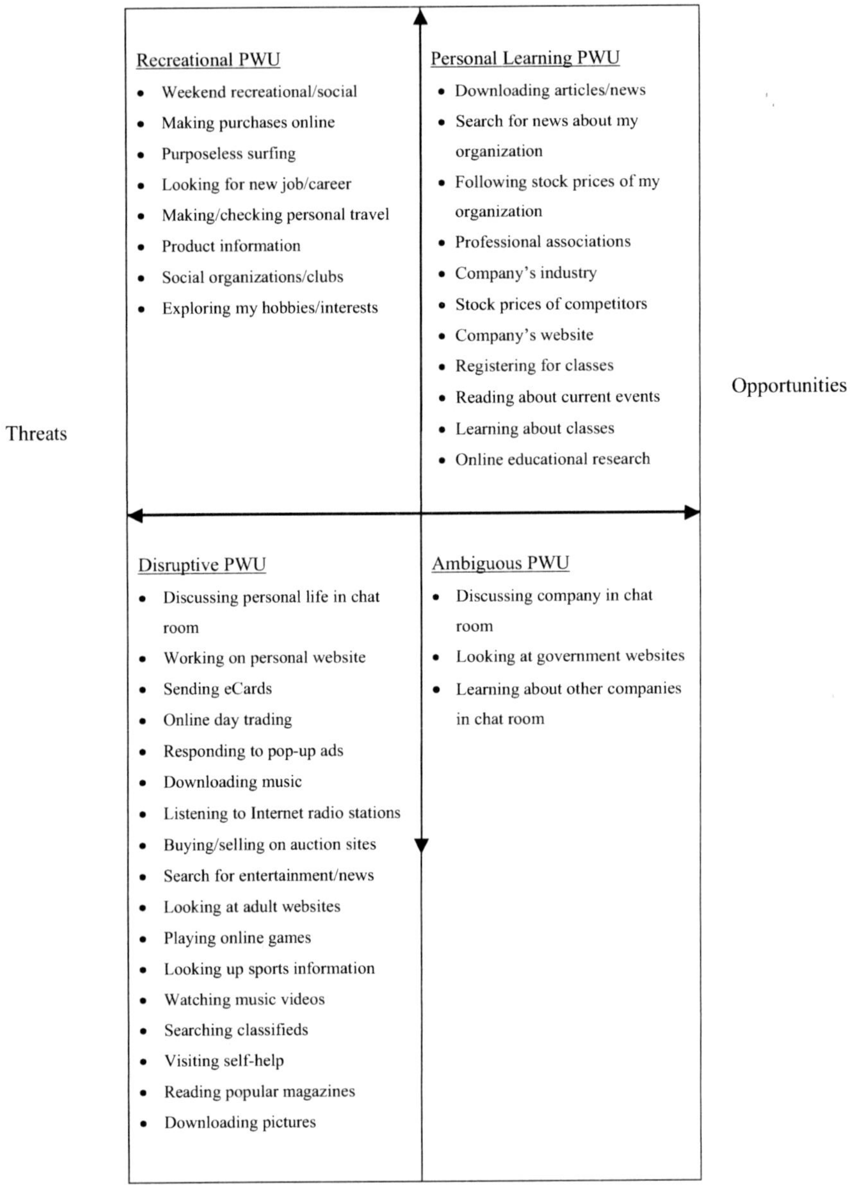

performed which focused on numerous aspects of this phenomenon and attempted to explain it using prior existing frameworks. The current study takes a contrary approach, letting the data itself propose the framework in an effort to truly understand how workers feel about PWU behaviors. It is hoped that by deriving a typology in such a fashion, researchers will have a more accurate foundation from which to pursue courses of study, and management and employees alike will be better able to maximize the benefits derived from, and reduce the costs associated with, personal Web usage in the workplace.

Multidimensional scaling was utilized as the analytical tool, which would best allow the research goals to be accomplished. As detailed above, MDS is particularly suited to such a study as it allows the dimensions to be accurately formed from the respondents themselves. A possible downside of such a methodology is that no significant structure may be revealed at all. Yet, this limitation may also be viewed as a positive of such a study, as it prevents the researcher from imposing a structure on the data that does not exist, hence "making something out of nothing at all."

Having performed the study, the data demonstrated that PWU may be reflected in specific behaviors that lie along two distinct axes. The first is based on productivity, with behaviors lying on a continuum ranging from threats to opportunities. Phrased differently, this implies that personal Web usage may be assessed in terms of whether it poses a risk or has the potential for constructive contributions. This is notable for two specific reasons. While the fact that certain behaviors can have negative consequences may not surprise managers, it is notable that the employees themselves not only recognize this fact, but also assess their PWU behavior along this line of thought. The second interesting result of this finding is that many workers find that certain personal Web usage behaviors are perceived to present opportunities for positive outcomes, or at the very least non-negative outcomes, such as no loss of productivity, no waste of time, and learning. This result implies that PWU may not necessarily be the drain on companies' resources that it is often purported to be. Indeed, developing research is exploring the possibility that certain forms of PWU may actually act as a catalyst, contributing to productivity in a myriad of ways ranging from reducing stress to increasing learning (Belanger & Van Slyke, 2002).

The second axis was based on the dimension of organization versus interpersonal. Specifically, it was viewed as employees' perceptions of the main entity affected by their PWU behaviors. Similar to the aforementioned continuum, it is significant that employees actually perceive their PWU as potentially detrimental to the company. The organizational/opportunity quadrant is ambiguous and needs further study, as there were insufficient behaviors in this cluster to confidently make conclusions.

This study has shown that when employees assess their PWU behaviors, they do so along two continuums. There are implications for both practitioners and researchers. From a management standpoint, this work may affect everything from how employees are initially trained to dictating certain Internet usage policies. A blanket policy prohibiting all PWU may indeed be "throwing the baby out with the bathwater." As the line between work life and home life becomes increasingly murky, this point may be all the more significant. Management may come to realize that 15 minutes of bill paying, game playing, and car buying online may be far better than eight to 10 hours lost due to sick days and personal time taken to accomplish the same tasks.

From a research perspective, this analysis is only an important first step, and a first step not without its limitations. The study itself is derived from a small, geographically concise sample, and replication using other samples from other geographically diverse locations would strengthen the validity of the results. Additionally, this study assesses employees' perceptions about their PWU behaviors; whether these behaviors actually do affect productivity and harm to the company, and in what fashion, remains to be empirically tested. As stated earlier, we have only just begun to look at this new area of technology and business, but the results derived thus far encourage further exploration.

REFERENCES

Anandarajan, M. (2002). Internet abuse in the workplace. *Communications of the ACM,* 45(1), 53-54.

Anandarajan, M. (2002). Profiling Web usage in the workplace: A behavior-based artificial intelligence approach. *Journal of Management Information Systems,* 19(1), 243-266.

Anandarajan, M., Simmers, C.A., & Igbaria, M. (2000). An exploratory investigation of the antecedents and impact of Internet usage: An individual perspective. *Behavior & Information Technology,* 19(1), 69-85.

Armstrong, L., Phillips, J.G., & Saling, L.L. (2000). Potential determinants of heavier Internet usage. *International Journal of Human-Computer Studies,* 53, 537-550.

Belanger, F. & Van Slyke, C. (2002). Abuse or learning? *Communications of the ACM,* 45(1), 64-65.

Computer Economics. (2002, January 4). Malicious code attacks had $13.2 billion economic impact in 2001. Available online at: http://www.computereconomics.com/article.cfm?id=133.

Elron Report. (2000). Available online at: www.elronsw.com.

Greengard, S. (2000). The high cost of cyberslacking. *Workforce,* 79(12), 22-24.

Griffiths, M. (1998). Internet addiction: Does it really exist? In Gackenbach, J. (Ed.), *Psychology and the Internet: Intrapersonal, Interpersonal, and Transpersonal Implications*. San Diego, CA: Academic Press.

Horrigan, J.B. (2002). Getting serious online. *Pew Internet & American Life Project,* (March 3).

Joinson, A. (1998). Causes and implications of disinhibited behavior on the Internet. In Gackenbach, J. (Ed.), *Psychology and the Internet: Intrapersonal, Interpersonal, and Transpersonal Implications*. San Diego, CA: Academic Press.

Korgaonkar, P.K. & Wolin, L.D. (2002). A multivariate analysis of Web usage. *Journal of Advertising Research,* 39(2), 53-68.

Kruskal, K.B. & Wish, M. (1978). Multi-dimensional scaling. Sage University paper series on *Quantitative Applications in the Social Sciences* 07-11. Beverly Hills and London: Sage.

Lederer, A.L., Maupin, D.J., Sena, M.P., & Zhuang, Y. (2000). The Technology Acceptance Model and the World Wide Web. *Decision Support Systems,* 29(3), 269-282.

Lim, V.K.G. (2002). The IT way of loafing on the job: Cyber loafing, neutralizing and organizational justice. *Organizational Behavior Journal,* 23(5), 675-694.

Lim, V.K.G., Teo, T.S.H., & Loo, G.L. (2002). How do I loaf here? Let me count the ways. *Communications of the ACM,* 45(1), 66-70.

Mills, J.E. (2001). Cyberslacking! A liability issue for wired workplaces. *Cornell Hotel and Restaurant Administration Quarterly,* 42(5), 34.

Oravec, J.A. (2002). Constructive approaches to Internet recreation in the workplace. *Communications of the ACM,* 45(1), 60-63.

O'Hare, D. (1976). Individual differences in perceived similarity and preference for visual art: A multidimensional scaling analysis. *Perception and Psychophysics,* 20, 445-452.

Panko, R.R. & Beh, H.G. (2002). Monitoring for pornography and sexual harassment. *Communications of the ACM,* 45(1), 84-87.

Poster, M. (2002). Workers as cyborgs: Labor and networked computers. *Journal of Labor Research,* 23(3), 339.

Punj, G. & Steward, D.W. (1983). Cluster analysis in marketing research: Review and suggestions for applications. *Journal of Marketing Research,* 20, 134-148.

Putnam, D.E. & Maheu, M.M. (2000). Online sexual addiction and compulsivity: Integrating Web resources and behavioral telehealth in treatment. *Sexual Addiction & Compulsivity,* 7, 91-112.

Roberts, B. (1999). Filtering software blocks employees' Web abuses. *HR Magazine,* 44(9), 114-120.

Rushton J.P., Brainerd, C.J., & Pressley, M. (1983). Behavioral development and construct validity: The principle of aggregation. *Psychological Bulletin,* 94, 18-38.

Schiffman, S., Reynolds, M.L., & Forest, Y. (1981). *Introduction to Multidimensional Scaling: Theory, Methods and Applications.* London: Academic Press.

Siau, K. & Nah, F.F.H. (2002). Acceptable Internet use policy. *Communications of the ACM,* 45(1), 75-79.

Simmers, C.A. (2002). Aligning Internet usage with business priorities. *Communications of the ACM,* 45(1), 71-74.

Simmers, C.A. & Anandarajan, M. (2002). Perceptions of personal Web usage in the workplace: A concept mapping approach. *Academy of Management Conference.*

Sloan, P. (2002). The temptations of the Web. *Database and Network Journal,* 32(4), 11-12.

Surf Control Report. (2000). *Surfing the Web at Work: Corporate Networks are Paying the Price.* White paper, (July 1).

Teo, T.S.H. & Lim, V.K.G. (1998). Usage and perceptions of the Internet: What has age got to do with it? *Cyber Psychology & Behavior,* 1(4), 371-381.

Teo, T.S.H., Lim, V.K.G., & Lai, R.Y.C. (1999). Intrinsic and extrinsic motivation in Internet usage. *Omega,* 27, 25-37.

Thompson, P. (1983). Some missing data patterns for multidimensional scaling. *Applied Psychological Measurement,* 7, 45-55.

Urbaczewski, A. & Jessup, L.M. (2002). Does electronic monitoring of employee Internet usage work? *Communications of the ACM,* 45(1), 80-83.

U.S. Department of Commerce. (2002). *A Nation Online: How Americans Are Expanding Their Use of the Internet.* Available online at: http://www.ntia.doc.gov/ntiahome/dn/index.html.

Websense. (2002). *Web@work Survey 2002: Cyber-Addiction in the Workplace.*

APPENDIX
LIST OF PWU BEHAVIORS

Code	PWU Behavior
PWU 1	Discussing personal life in a chat room
PWU 2	Working on my own personal website
PWU 3	Searching for weekend recreational/social activities
PWU 4	Downloading articles/news
PWU 5	Making purchases online
PWU 6	Discussing my company in a chat room
PWU 7	Sending eCards and post cards
PWU 8	Searching for news about my organization
PWU 9	Purposeless surfing of the web
PWU 10	Online day trading
PWU 11	Responding to pop-up advertisements
PWU 12	Downloading music
PWU 13	Looking for a new job/career
PWU 14	Making/checking personal travel arrangements
PWU 15	Listening to Internet radio stations
PWU 16	Following stock prices for my company
PWU 17	Buying/selling objects from online auctions such as eBay
PWU 18	Downloading pictures
PWU 19	Looking at government websites
PWU 20	Finding information about products I wish to purchase
PWU 21	Visiting professional associations' websites
PWU 22	Reading about my company's industry
PWU 23	Exploring my hobbies/interests
PWU 24	Searching for entertainment news
PWU 25	Following stock of competitors
PWU 26	Looking at adult websites
PWU 27	Reviewing my company's website
PWU 28	Playing online games
PWU 29	Registering for educational/training classes
PWU 30	Learning about other companies in a chat room
PWU 31	Reading about current events

(continued on the following page)

(continued from the previous page)

PWU 32	Looking up sports information (scores, statistics, team info)
PWU 33	Learning about educational/training classes
PWU 34	Visiting websites of my social organizations/clubs
PWU 35	Watching music videos/Internet movies
PWU 36	Online research for educational purposes
PWU 37	Searching online classified ads for apartments/real estate
PWU 38	Visiting self-help websites
PWU 39	Reading online versions of popular magazines/newspapers

Email, though employing the medium of the Internet, does not necessitate the use of the World Wide Web and is thus not encompassed by this definition for this study

Section II

Managing Personal Web Usage from a Human Resource Perspective

Chapter V

The Effect of Trust on Personal Web Usage in the Workplace

Susan K. Lippert
Drexel University, USA

ABSTRACT

This chapter addresses the concept and importance of interpersonal trust through the use of the Internet in an organizational setting. In particular, personal Web usage is explored by examining employee interpersonal trust. Personal Web use refers to an employee's utilization of the Internet for non-job related activities within a work environment. Examples of personal Web use include online banking, participating in instant messaging or chat sessions, buying goods or services, and any other activity in which the Internet is accessed for non-work-related tasks. A discussion regarding the importance of trust, its nature, and strategies for building interpersonal trust in an organizational setting are offered. Generalized guidelines for organizational practice and recommendations to support a culture of trust within the work environment are presented. This chapter addresses the notion of trust through personal Web usage as a human resource management issue.

WHY IS TRUST IMPORTANT?

Trust is important in organizations due to the potential economic savings derived from increasing trust between individuals (Williamson, 1975). There is an inverse relationship between transaction costs and trust, such that as trust increases, costs decrease (Bromiley & Cummings, 1995). Transaction costs are expenditures for controlling, monitoring, and processing work-related activities. Processing costs, a subset of transaction costs, include the extrinsic and intrinsic costs of doing business, both in staff and line functions. By developing trust, a company can benefit through lower processing costs — a bottom-line outcome. Trust, as defined in this section, is the "individual's belief or a common belief among a group of individuals that another individual or group: (1) makes good-faith efforts to behave in accordance with any commitments both implicit and explicit; (2) is honest in whatever negotiation preceded the commitments; and (3) does not take excessive advantage of another even when the opportunity is available" (Cummings & Bromiley, 1996, p. 303).

Trust is the 'glue' that holds everything in society together (Luhmann, 1979) and is an important element of human relations (Larzelere & Huston, 1980). Trust is central to transactions (Dasgupta, 1988), because without trust, we are frequently immobilized through an inability to make a prediction or fulfill expectations. Trust can be used as an indicator of individual, group, organizational, or cultural health since the entity of trust can be a person, place, event, or object (Giffin, 1967). Trust can exist between individuals and organizations (Zaheer, McEvily, & Perrone, 1998), between organizations (Gulati, 1995), between users and information technology (Lippert, 2001), as a general characteristic of different societies (Fukuyama, 1995), or as an interpersonal exchange between individuals (Mayer, Davis, & Schoorman, 1995). In organizations with high levels of trust, productivity consistently exceeds other businesses where trust is low or latent (Sitkin & Stickel, 1995; Davis, Mayer, & Schoorman, 1995). *Trust is a measure of the effective interaction between individuals*. The development of interpersonal trust relationships is important for sustaining individual and organizational effectiveness (McAllister, 1995). Trust permits us to have some degree of predictability of another's behaviors which allows us to establish and test expectations (Deutsch, 1958, 1960). The ability to test expectations enables us to develop and maintain social order (Arrow, 1974).

PERSONAL WEB USAGE AND TRUST

Interpersonal trust can be used to monitor, measure, and ultimately influence personal Web usage in an organizational environment. The link between trust and Web use exists through the degree to which an individual trusts the organization in which she is employed. The use of the Internet for personal activities can and will manifest through trust behavior. Trust is a metric for measuring Internet usage and serves as a proxy for functional or dysfunctional use. Functional Web behavior can be defined as the degree of Internet use to conduct personal business during work hours that conforms to and follows organizational policy. Personal Web use is presently controlled through organizational rules, regulations, policy, and actions. In an organizational context, policy is established, worker behavior is observed, and subsequent transgressions are addressed.

Dysfunctional behavior constitutes a misappropriation of organizational time and resources that would not otherwise be sanctioned by co-workers or supervisors. Through measuring interpersonal trust and through the development of increased levels of trust, dysfunctional Web use can be discouraged. In this chapter, the development of trust is examined as an alternative strategy to increase appropriate personal Web use behavior.

In a trust-rich organizational culture, daily employee Internet activities would occur in accordance with established written protocols so that Internet access and usage is appropriate in *all* transactions. The organization and its leaders would maintain a fundamental respect for employees as well as each other. Internal business interactions between employees would be conducted consciously and consistently in all activities. The need for overt control mechanisms to monitor employee behavior diminishes. Employees are treated as vested partners, and enlightened leadership exists through examples of trustworthy behavior by all levels of management. Work environments where trust can be explicated and discussed are valued. The maintenance of an employee's trustworthy reputation internally and externally is sought. Written procedures for dealing with trust breaches are established by management and when required acted upon quickly and publicly. The acceptance of different perspectives and an environment where feedback occurs is encouraged.

These notions represent a difference in perspective, an acceptance of the importance of trust, and a commitment to work directly and indirectly toward developing and maintaining a climate of trust. The consequence of this perspective is that the organization can ultimately be observed, measured, and described as having a 'trust culture.'

Building a trust culture is not easy, and creating cultural change is a long and arduous course of action. The process by which an organization develops and sustains an atmosphere of trust begins with taking the risk that employees are trustworthy. This becomes a starting organizational precept that is tested over time. What this means is that organizational leaders begin from a belief position that trustworthy behavior is the norm in the company, and set an example through their own trustworthy action. There is an implicit expectation that trust will exist in all interactions and that individuals who work for the company will act in a trustworthy manner.

The organization openly communicates about the nature of trust and this fundamental originating belief. In fact, creating and sustaining a culture of organizational trust becomes an overall long-term goal. Trust is explicitly addressed in the corporate values, the overall mission statement, and in specific employee functions. A measure of trust is created for annual performance reviews. Trust testing is done passively as individuals interact with one another. Trust breach assessments are limited to the specific incident (Robinson, 1996), and the corresponding effect and response remain isolated rather than generalized beyond the situation. Periodic assessments of the perception of trust in the corporate environment are undertaken. Understanding the importance of trust, what constitutes trust, and how to build and develop trust provides a basis for enhancing organizational interactions and engenders a process for individual development.

THE NATURE OF TRUST

The notion of trust is often used in daily conversation. Many times people make comments such as: "I trust my supervisor because she is a friend" or "She has always been honest with me." A trust relationship occurs when one individual (the trustor) can or does trust another individual (the trustee). This relationship develops through a series of interactions between two entities over time (Rempel, Holmes, & Zanna, 1985; Zand, 1972). Reliance upon information received from another person about uncertain situations and their potential outcomes produce a possibility of risk (Schlenker, Helm, & Tedeschi, 1973). Blau (1968) suggests that trust relationships build slowly starting with transactions requiring limited risk, enabling both the trustor and the trustee to demonstrate their trustworthiness.

The relationship between the trustor and the trustee matures as a function of repeated trust assessments associated with subsequent interactions. It has been said that a person consciously or unconsciously evaluates every situation and decides if an individual is worthy of greater or lesser trust.

Individuals, within trust relationships, are evaluated based on an expectation of how a person will react, behave, or function in a given situation (Zucker, 1986). McGregor (1967) suggests that inconsistencies between explicated thoughts and actions serve to decrease trust. The resultant trust level is contingent upon the consistency of behavior over time and across interactions. In order for trust to develop and be sustained, the individual's actions must be predictable with some degree of accuracy. Rempel and Holmes (1986) assert that a person is said to be predictable if their behavior is consistent. An expected action may result in either a positive or negative outcome (Mishra, 1993). While we may hope for a specific result, we can still sustain trust development as long as what occurs is congruent with what we expect to happen (Barber, 1983).

Trust functions through the bi-directional relationship between individuals. In one direction, there is trust from the employee to his manager and there is also trust from the manager to the employee. Therefore, both entities concurrently take on the role of trustor and trustee in this dyadic relationship. Figure 1 depicts this bi-directional relationship.

Trust is generally contextual (McKnight, Cummings, & Chervany, 1996, 1998; McKnight & Chervany, 1996). However, we all have what Rotter (1967) calls, a "predisposition to trust." Predisposition to trust means an

Figure 1. Bi-Directional Relationship Between Two Individuals

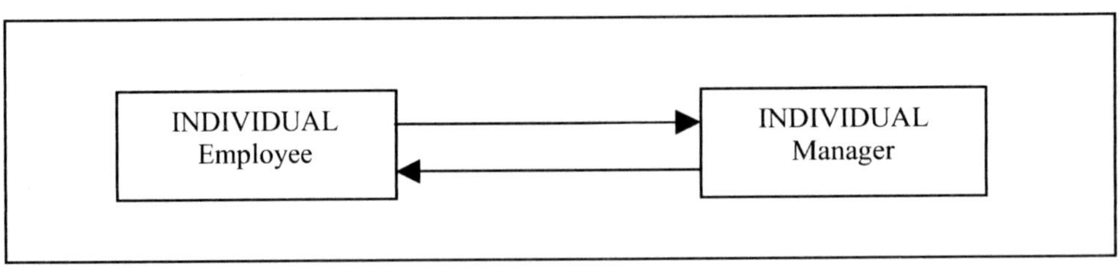

orientation, based on past experiences, to be more or less trusting of others (Rotter, 1971). Each person will have a different level of predisposition to trust based on his/her past experiences. In order for trust to exist, past experiences are needed to establish familiarity with the situation (Luhmann, 1979). We observe our world from the time we are children, and with each new experience we add to our personal database of what constitutes acceptable and unacceptable conduct. Over time, we develop a predisposition to trust at some level and apply this to a specific set of conditions or contexts. By the time we are adults, we have a set of tacit beliefs, which when applied to our environment, both workplace and other, leads to an increased probability of being able to predict an outcome — our level of trust.

Predisposition to trust has two forms. The first type of predisposition is based on the sum of all life experiences and is called *general predisposition*. Each interaction throughout the course of an individual's life adds to her perception of a general sense of trustworthiness within society. A geographic group might classify themselves as skeptical or suspicious in business transactions. Within this example, an individual's predisposition to trust in a business transaction might be described as low regardless of the referent group. A referent group is a collection of individuals who are linked in some way — through business, ideology, interest, geographic region, or even gender—and who share a set of common characteristics (Hogg & Terry, 2000).

The second type of predisposition to trust is referent group specific and is called *contextual predisposition*. The trustor's predisposition changes over time based on past experiences. For example, if in the past an individual's interactions with his previous managers have been positive, he will likely have a high predisposition to trust his managers in the future. In this case, the predisposition to trust is the sum of the experiences associated with the specific referent group — the managers. General and contextual predispositions, when joined, form a combined predisposition to trust.

Trust is a perception crafted by the trustor about a particular person within a specific situation (Gabarro, 1987). Trust is also a mental state which changes as additional data are collected. Every interaction is evaluated and judged by the trustor and the trustee. Trust levels are affected by the degree of vulnerability experienced by the trustor (Mayer, Davis, & Schoorman, 1995). For low levels of vulnerability, where the actions of the trustee result in a minimal risk, the level of trust diminishes slightly. However, if a trustee fails to perform an action which places the trustor at a significant risk, a greater level of trust loss results (Deutsch, 1973). The willingness to take risks may be a common characteristic of all trust situations (Johnson-George & Swap, 1982). For example, if an employee tells a co-worker that he needs to complete an Internet-based task but instead is found to be conducting personal banking activities, the co-worker will re-evaluate the situation and reassess the reliability of the employee. This is a relatively low-risk, low-vulnerability scenario. If, however, the employee is able to provide an adequate explanation such as having been traveling for work for the past several days without adequate time to make the mortgage payment, the co-worker may understand and accept the rationale, resulting in minimal or no loss of trust. If, on the other hand, the employee fails to offer an adequate justification or is engaged in a morally indefensible online transaction, the co-worker will begin to lower his estimation of the employee. In this situation, the degree of risk felt by the co-worker is negligible and therefore the trust violation may be trivial.

In a different example, a co-worker promises to electronically submit time cards for a sick colleague. However, the co-worker fails to meet the deadline for submission because he used the available time to conduct personal business on the Internet. The missed deadline resulted in a delayed paycheck for the sick colleague. The magnitude of the trust expectation is relatively high since the colleague needed the money in order to pay her bills. As such, she experiences moderate to high risk and vulnerability and the resultant level of trust is significantly affected. The inattentive action of the co-worker significantly lowers the colleague's trust level.

Every situation affects the overall assessment of trust between two individuals in a different way. Trust exists on a continuum from low to high trust as depicted in Figure 2.

The trustor can place the trustee at any point along this continuum. Each subsequent interaction will shift the overall trust evaluation either to the right or

Figure 2. A Trust (Intensity) Continuum

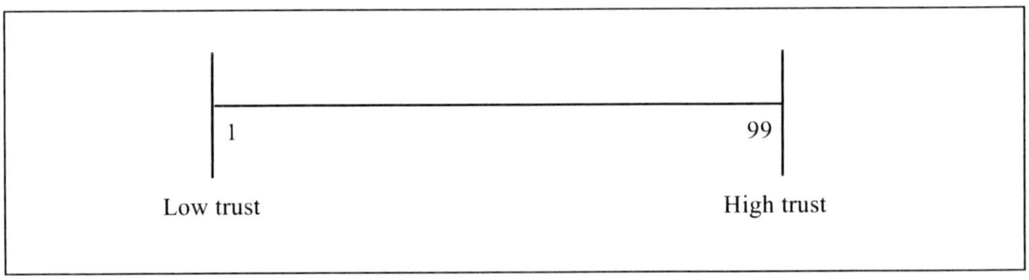

left on the continuum. The degree of trust will vary based on: (1) the contextual environment, such as the workplace or a personal or social setting; (2) an individual's predisposition to trust; (3) the magnitude of the interaction; (4) the current state of the trust relationship; and (5) the time since the last significant trust interaction. We can even label the trust interaction phenomenon as a Significant Trust Event (STE). An STE is any trust interaction that has a significant effect on the resultant trust level.

Figure 3 graphically depicts three independent sequential interactions between two individuals. In the first interaction (a), the trustor evaluates an experience resulting in a low-trust assessment. At the end of the interaction, the resultant or current state of the trust relationship (b) is relatively low. The second interaction (c) is one of great importance (magnitude–higher STE) to the trustor which results in a higher trust assessment since the trustor engaged

Figure 3. Levels of Trust

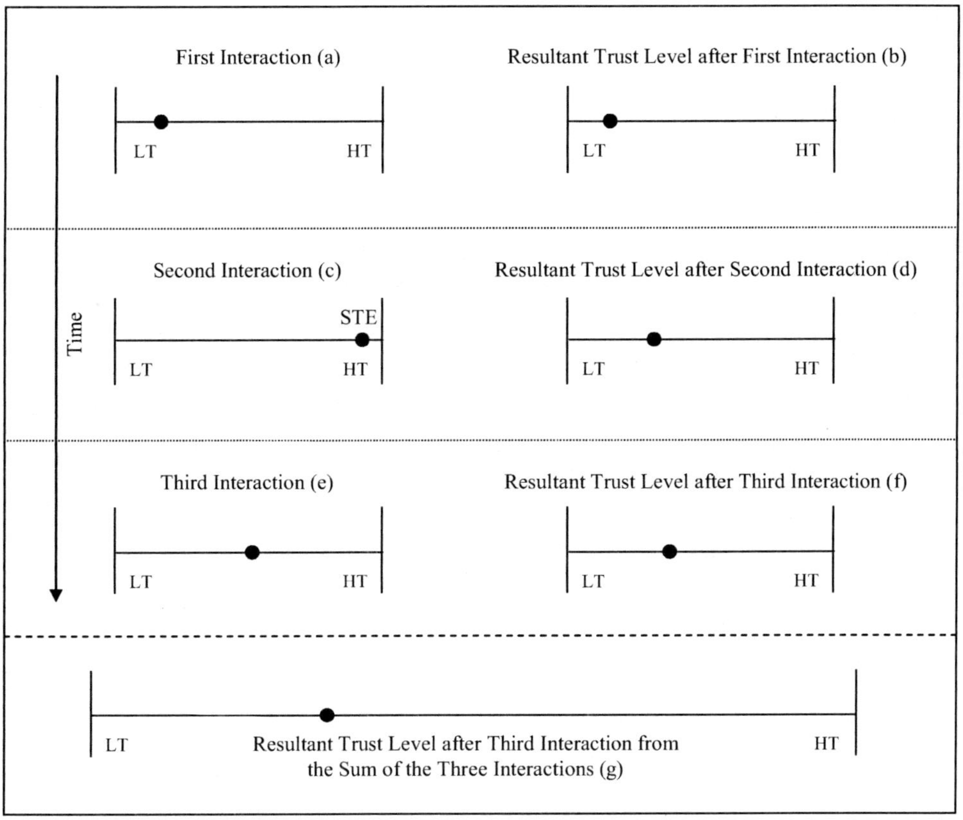

in a positive and important interaction. The resultant state (d) of the trust relationship increases to a greater degree after the second interaction due to the positive outcome and magnitude. The third interaction (e) was of limited importance (magnitude) to the trustor, and the assessment of that interaction is a slight diminishment (or loss) of trust. The level of trust resulting from the third interaction (f) between these same two individuals is stronger than the first interaction but lower than the second interaction. Each of these three independent interactions resulted in an overall level of trust for the trustor of the trustee (g). As long as the interaction or events occur between the two individuals, the level of trust will move back and forth based on the sum of all their interactions. Therefore, the resultant trust level at any point in time is the sum of the sequenced events. As each new interaction occurs, the overall trust assessment will continue to shift along the trust continuum.

Trust, in this context, is a variable and lies along the trust continuum. Trust varies and changes with each subsequent interaction. In every interaction between two people, a judgment is made that affects the overall evaluation of trust—the resultant trust level. The magnitude which a person places on an exchange is determined based on the significance of that interaction — the extent of the significant trust event. A value judgment is made which determines the importance of the transaction which is then factored into the totality of all other transactions in order to determine the value of the judgment. Trust becomes the outcome state placed upon the trustee. It should be fairly evident that one of the difficulties with the notion of trust is that the word can be used as a noun, a verb, or an adjective resulting in slightly different connotations. Defining trust is often considered problematic due to the wide variance of meaning (Hosmer, 1995).

Trust can be transitory or short-lived (Lippert, 2002). The degree of vulnerability the trustor feels will impact the fleeting nature of trust. Trust is also a temporary end state. At the end of several interactions, the trustor makes a determination of the trustworthiness of the trustee. The cumulative experiences (Gabarro, 1987) which establish the trustworthiness of the individual on a continual basis are summed to the end state of saying that an individual can be 'trusted.'

BUILDING TRUST IN ORGANIZATIONS

Building trust between individuals within organizations is accomplished through a series of sequential phases. Lewicki and Bunker (1996) offer a model based on the work of Boon and Holmes (1991) and Shapiro, Sheppard, and Cheraskin (1992) which suggests that trust relationships move through three developmental stages — calculative-based trust, knowledge-based trust, and identification-based trust. It is important to understand what needs to occur in each stage of trust development in order to effectively increase the level of trust between individuals.

Calculative-based trust is a stage where each potential interaction between two individuals is assessed as an independent value-based transaction (Coleman, 1990). If the interaction is evaluated as beneficial to the trustor, he will engage in the transaction with the trustee. Every interaction is calculated to determine its potential value (Gambetta, 1988) and if a positive outcome is forecast, the trust level increases incrementally based on the perceived magnitude of the transaction. If the interaction outcome is negative, the trust relationship is diminished proportional to the scale of the violation. The value or weight of each transaction is compared to the outcomes associated with maintaining the relationship (Lewicki & Bunker, 1995; Shapiro, Sheppard, & Cheraskin, 1992).

In a calculative-based interaction, an individual can behave out of a concern for retribution (deterrence) for not following through on an obligation. Trust is sustained through the threat of punishment which motivates the trustee to a greater degree than the prospective of reward. Calculative-based trust, however, is quite tenuous and is highly susceptible to extinction of the relationship based on a single flagrant action. In situations where the magnitude of the action is egregious, the trustor can 'calculate' that the relationship should not be sustained. Therefore, in calculative-based trust, the trust relationship can be completely severed if the trustor feels that the magnitude of the action is severe.

Knowledge-based trust is grounded in an individual's degree of predictability. If the trustor can predict with a relatively high degree of certainty how the trustee will behave, the trust relationship will continue to grow. When behaviors can be anticipated, a degree of generalized expectancy occurs. The predictability of behavior, over time, derived from the accumulation of knowledge through experience with the other person, enhances trust (Holmes, 1991).

Two key processes are necessary to build trust in the knowledge-based trust phase. The first process, *explicit communication*, enables the parties to express their thoughts, concerns, and expectations openly and honestly. Explicit communication entails the use of verbal and non-verbal mechanisms necessary to establish a common understanding and achieve shared knowledge between the two parties. The second process, *nurturing*, involves a stylized set of behaviors necessary to establish a richer connection and compatibility between the individuals. During the second process, the trustor continues to watch and listen to the trustee with whom he engages in explicit communication. This encourages the trust relationship. Relationships within an organizational context are often knowledge-based. Trust, at the knowledge-based level, is minimally affected by inconsistent behavior. If the trustee can adequately explain the reason for his behavior, the trustor is likely to accept the justification with little to no impact on the resultant trust level.

Identification-based trust is the third phase of a trust relationship. In this stage, the trustor and trustee can effectively understand and appreciate the other's needs. This permits the trustor to function as the trustee's agent. In this stage of trust development, both parties learn what really matters to each other, thus enabling them to eventually place the same degree of importance on those behaviors. In this stage, the individuals are able to understand one another without the need for protracted explicated conversations. The trustor and trustee are synchronized in understanding what is important to each other. Both individuals work consciously to be supportive of the other and are respectful of the other's concerns. Very few relationships reach this stage of trust in an organizational setting because individuals often lack the time, energy, or interest necessary to achieve this highest level of trust. Figure 4, adopted from Lewicki and Bunker (1996), depicts the three phases of trust development. The curve represents the development of trust through the three stages over time.

Trust exists in a business relationship when three conditions are met: (1) the parties risk losing too much if either individual behaves inappropriately; (2)

Figure 4. Stages of Trust Development in the Work Setting (Adopted from Lewicki & Bunker, 1996)

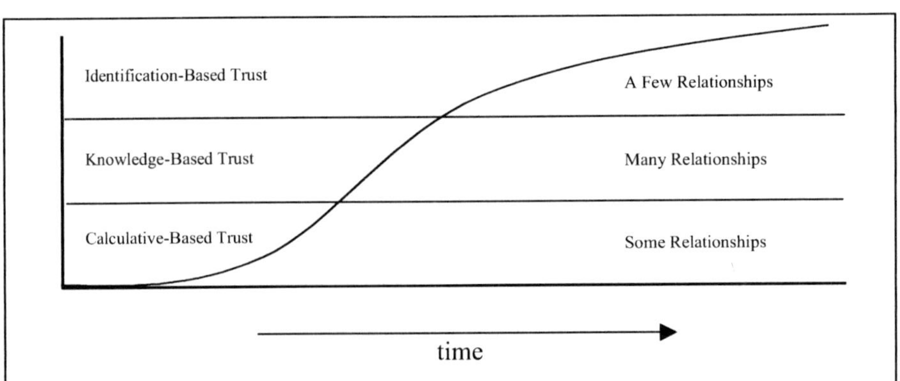

either individual can predict the other's behavior well and can therefore protect against being deceived; and (3) both individuals have adopted the other's preferences (Sheppard & Tuchinsky, 1996, p. 143). Although trust is difficult to build (Tyler & Degoey, 1995), developing trust within organizations is facilitated through meeting conditions, understanding stages, and taking explicit actions consistent with the trust relationship phases.

EXAMPLES OF PERSONAL WEB USAGE

There are many ways that trust is manifest in the use of the Web for personal business in the workplace. While not exhaustive, Table 1 offers

Table 1. Examples of Personal Web Usage

Area of Association	Activity
Personal Finances	◆ paying bills through an online billing paying service ◆ adding funds to a telephone card used solely for personal use
Mortgage	◆ checking for updated mortgage rates before contacting a vendor online about refinancing a home mortgage
Travel	◆ making travel plans for the family through an online site ◆ updating travel preferences with various online airlines, train or bus facilities
Family Activities	◆ researching information for by the parents for a child's class project ◆ investigating weekend activities which might be fun for the family ◆ checking a personal e-mail account, and staying in contact with friends and family during the workday
Friend Activities	◆ e-mailing a friend about dinner through a corporate e-mail account ◆ searching for restaurants where you might go for dinner ◆ engaging in an instant message session with a friend ◆ participating in a chat room discussion with another individual who may be a friend, family member, or colleague ◆ sending photographs to friends and relatives ◆ playing solitaire or games with a group of individuals online
Searching	◆ looking for downloadable software related to personal activities such as managing photographs, additional calculator functions, or applets for the palm pilot ◆ locating a telephone number ◆ searching for an old acquaintance ◆ perusing graphic photographs which others might consider offensive or even pornographic
Personals	◆ looking up potential personals through an online search capability ◆ communicating with individuals identified through a personals online search capability
Purchasing	◆ investigating the qualities of a product in preparation for a potential purchase such as researching automotive options or dealers ◆ making plans to purchases goods such as groceries, a computer, CDs, books, or other personal items
Personal Web Page	◆ posting pictures to a personal Web page ◆ developing text for inclusion on a personal Web page

examples of Web usage which fall within the domain of non-work-associated activities.

There is variance about what constitutes acceptable and unacceptable Web usage in the work environment. This difference is based on the individual's applied ethic, the standards of acceptability of the referent group, and the organizational policies. While most of the examples in the above table might be viewed as insignificant, there are some uses of the Web that are commonly accepted as inappropriate regardless of differences in personal preferences or point of view. Downloading illegal, immoral, unethical, or distasteful material is a common breach of trust that is unacceptable regardless of context or condition.

When making a judgment about the acceptability of marginal activities in workplace Web usage, the ethics of referent groups can serve as a guide. A good test of appropriate personal Web usage is whether an individual would feel comfortable doing personal Web functions with the knowledge of their supervisor or a family member. An initial reference baseline establishing an acceptable threshold of behavior is necessary. This can be accomplished through the use of organizational policies, referent group common practice, or what constitutes socially acceptable behavior. These codes of behavior establish a foundation for refining a set of practices that permit the development of acceptable Web usage by individuals within the workplace.

INDIVIDUAL TRUST BUILDING WITHIN ORGANIZATIONS

Some initial recommendations for building individual trust relationships between employee members emerge from the Lewicki and Bunker (1996) model. Since relationships within organizations rarely move beyond the knowledge-based trust level, it is important to understand how to effectively build and maintain trust at both the calculative-based and knowledge-based levels.

In order to build trust with a new employee or to sustain trust in a relationship at the calculative-based level, it is important that both parties behave consistently in their interactions over time. The issue of consistency and congruity of behavior and action is a principle factor in building trust at the calculative-based stage. Within an organizational setting, managers need to establish deterrence measures which withdraw benefits from the trustee if he behaves in an untrustworthy manner. Influence of the trustee's behavior through the potential loss of future interactions or impact on his reputation with co-workers must outweigh the potential benefit derived from behaving in an untrustworthy fashion. A temporary gain associated with an untrustworthy action needs to be offset by the enduring advantage of preserving a high-quality reputation. Deterrence of untrustworthy actions requires monitoring and oversight between individuals to ensure that trust violations do not occur.

An example of an interim gain might be a situation where an employee, on company time, downloads software for personal use onto a work computer, after company policy has been disseminated prohibiting such activity. The employee chooses to download the software regardless of the previous admonishment. In this case, the short-term gain of accomplishing the download task is offset if the employee is caught in violation of company policy. The choice, to violate or not, is an employee action that affects trust. If the violation is identified, for example by a co-worker who happens by, the long-term impact on the employee's integrity and trustworthiness can be significant. While disciplinary actions could be implemented to match the policy violation, the greater effect is on the employee's reputation. In this case, trust is degraded and the employee is viewed with skepticism since his trustworthiness has diminished. The magnitude of the trust violation or trust action oftentimes only has an effect on the referent group. In other words, the trust relationship between the employee and a friend in another company would not be impacted by the violation. The severity of the trust building or diminishing behavior is directly proportional to the assessed trust level within the referent group.

To continue building trust, all parties must consciously consider and be explicitly clear about their intentions both prior and during interactions (Mellinger, 1956). To help accomplish this clarity, individuals must state their expectations, describe their reasoning, and offer explanations associated with their intent. To verify this shared understanding, either party can seek clarification through feedback and discussion (Argyris, 1965; Argyris & Schöen, 1978; Argyris, 1990). Many disagreements can be resolved by discussing the situation in an open (Butler, 1991) and non-defensive manner.

For relationships functioning at the knowledge-based trust stage, parties build trust through consistently congruent behavior. Individuals make trust assessments associated with a knowledge-based trust relationship and utilize information (knowledge) from past interactions. The cumulative interactions between the two individuals over time provide a basis for knowing with a degree of certainty how the other will behave in a specific situation. This sum of all interactions enables the trustor to develop a generalized expectancy of how the other will behave.

Information contributes to the predictability of the trustee, and predictability enhances trust. Accurate estimation of another's behavior requires information which the trustor collects and evaluates through repeated interactions. In order to develop knowledge-based trust, the individuals need to continually communicate with one another with an earnest interest in learning more about each other (Argyris & Schöen, 1978). Effective communication includes sharing information about concerns, wants, and inclinations, and learning occurs through observation of the other party. Data collected in different contextual situations allow the trustor to develop a broader trust perspective, since trust assessments are conducted under a variety of circumstances. This collection of information enables the trustor to predict how the trustee will behave. Calculative-based trust places a greater emphasis on a predisposition to trust because of the lack of historical data on which to base a trust judgment. Knowledge-based trust, however, uses information obtained from past interactions as the core data to make assessments.

In the development of trust, individuals can make a judgment through conscious reflection on a variety of different interactions. Table 2 provides some reflective questions which can be used to assess interpersonal trust at the knowledge-based level.

These questions, while not exclusive, provide a mechanism for measuring the trust relationship between individuals within an organizational context. They are most useful as a systematic and organized set of assays to assist in making a judgment about the trust level between organizational entities. The questions are offered as a starting guideline for evaluating a trust relationship between two organizational members; however, other areas of consideration for the trust relationship might be relevant.

THE IMPORTANCE OF ESTABLISHING AND MAINTAINING A CULTURE OF TRUST WITHIN ORGANIZATIONS

Establishing trust is more than simply developing trust in interpersonal relationships or between individuals and an organization. Organizations have oftentimes avoided the whole issue of organizational trust because of its

Table 2. Assays for Measuring the Trust Relationship for Individuals Within a Business Setting (Adopted from Sheppard & Tuchinsky, 1996, p. 146)

#	Question
1	I know my manager or co-worker will consider my concerns when making decisions.
2	The quality of our communications is extremely good.
3	We confront issues effectively.
4	We discuss the critical issues of our relationship well.
5	We have frequent face-to-face contact.
6	We speak frequently on the telephone.
7	We have a long history.
8	I expect to interact with my manager or co-worker for a long time in the future.
9	Our contacts entail many different issues. (Our relationship is multidimensional [Butler & Cantrell, 1984].)
10	Our goals are similar.
11	We have similar world views.
12	We are compensated for accomplishing the same outcomes.
13	I frequently think of my manager or co-worker as a member of the same organization (family).
14	We have many shared activities.
15	I know well the people important to my manager or co-worker.
16	My manager or co-worker knows well the people important to me.
17	I understand well the basis on which my manager or co-worker is rewarded and compensated.
18	My manager or co-worker understands well the basis on which I am rewarded and compensated.
19	I understand my manager's or co-worker's primary problems at work.
20	My manager or co-worker understands my problems at work.

complexity, its abstract nature, a lack of understanding of its importance, and because the actions to develop trust are generally illusive, vague, and at best, difficult to operationalize (Bhattacharya, Devinney & Pillutla, 1998). Organizations, for the most part, are just beginning to recognize the need for 'attention to trust' as a fundamental driving force behind organizational climate and, hence, the significant effect on the bottom line. Trust is an important operating force in any enterprise, and the explicit efforts to establish, nurture, and maintain a trust culture can make a considerable difference in business success.

Individuals can develop, nurture, pay attention to, and proactively work toward enhancing a culture of trust through consistent and trustworthy interactions with one another. The summative effect of individual conscious actions provides a model that influences the organizational culture. But how does an organization begin to institutionalize and habitualize trust behaviors and, more importantly, make the awareness of trust resolute throughout the working environment? One approach is through a strategic reframing of trust as an outcome of organizational processes. This idea of a culture of trust is consistent with the use of any organizational resources that are expended in the routine course of business. Organizations expend significant time, money, and talent to train employees, to provide updated equipment, and to streamline processes, in an attempt to optimize profit. An organization that actively promotes a culture of trust might be viewed as simply deploying yet a different resource to conduct business and develop competitive advantage. While there are some input costs to actively develop and maintain a climate of interpersonal trust, recent research into the effect of trust on the effectiveness and efficiency of business operations suggests that the overall cost is less than the benefit derived from establishing and maintaining a culture of trust (Sitkin & Stickel, 1995). Therefore, trust becomes a central core element of the corporate strategic goals rather than just another factor in the operational formula for organizational success.

In summary, the importance of trust is accepted, encouraged, and discussed through open, repeated, and consistent dialogue. A culture of organizational trust is explicated as a corporate goal. Specific and explicit actions are undertaken to nurture and maintain a climate of trust throughout the organization. Breaches of trust are handled with well-established and planned policy in a quick and public venue without generalization that could negatively impact the overall internal and external perception and reputation of organizational trust.

RECOMMENDATIONS FOR BUILDING TRUST

A series of conditions support trust building within organizations. These nine conditions, while tangent to the core development process, provide a framework for enhancing or degrading the trust-building activity. First, the consistency of each party's behavior is an important condition for successful trust development. Consistency of behavior is tied to reliability through the individual's confidence that a behavior (Cook & Wall, 1980), intention (Mellinger, 1956), need, want, or requirement is replicable. Consistency of action permits the trustor and trustee to predict that future interactions will follow expected patterns (Good, 1988).

Second, the conscious intelligence of each party shapes an individual's assessment protocol. Conscious intelligence refers to an individual's intent to mindfully conduct trust assessments and place the resultant trust evaluation along a perceptual continuum. Each assessment becomes the deliberately generated baseline for the next assessment and occurs concurrently with the sentient action. When conscious, the parties can knowingly apply the correct behaviors and the right amount of interaction for that stage. During the calculative-based trust stage, an individual's deliberate intention and resulting consistency of action over time becomes the impetus for movement into the knowledge-based trust stage. Within the knowledge-based trust phase, individuals can choose to intentionally explicate their feelings, wants, and needs, thus enabling the parties to determine the accuracy of the information and undertake corrective action if necessary. Regardless of the phase of trust development, individuals need to be open (Butler, 1991) to feedback and dialogue. Conscious intelligence is a required condition for enhancing individual trust building within organizations.

Third, the amount of time over which the relationship develops is a factor influencing the trust relationship. Interactions occur summatively and as they continue to occur, the amount of time the parties know each other concurrently increases. As time passes, the opportunity for repeated affirming interactions multiplies. The old adage that "time heals all wounds" is based on the idea that as time progresses, repeated actions affect individual perceptions, including the perception of trust. Time is a vital ingredient in the trust-building process because, over time, individuals have continuing interaction, which facilitates supplementary information and enables new opportunities for riposte.

Fourth, activities that weaken the trust-building process should be avoided. It is relatively easy to unconsciously neglect explicit trust-building behaviors and the support mechanisms that enhance trust development. As an example, two individuals are in a knowledge-based phase trust relationship. One or both of the parties allows their dialogue to become less important than other activities, thereby diminishing the quality of their communications. Because they are in a knowledge-based trust state, they assume that the other party is aware of their needs, wants, and intentions, and hence are not openly explicit about their position. The problem manifests because of an incorrect assumption of identification-based trust which does not match the needs or wants established during the knowledge-based trust phase. This has the effect of weakening the infrastructure that supports the trust relationship.

Fifth, develop an organizational climate or atmosphere that is receptive to trust building. Every organization can be classified on a climate scale ranging from a low trust to a high trust. Organizational climate is the collection of social indicators that provide a description of the working environment. Organizational climate includes how people interact with each other, how the organization is structured, the general working atmosphere, how people tend to 'feel' throughout the work day, and how the leadership affects motivation (Schneider, 1990). Climate is oftentimes described in the language of effect, such as happy, sad, stressed, competitive, lonely, or even Machiavellian. Some organizations do not have a climate that encourages trust development. Therefore in low-trust climate organizations, it is much more difficult to build and sustain trust than in those with more receptive environments. Managers can demonstrate a high level of trust for employees through delegation of risky tasks which will in turn lead to greater trust in the manger by the employee (Schoorman, Mayer, & Davis, 1996).

Sixth, trust has a cognitive and an emotional (affective) component (McAllister, 1995). We not only think about and assess trust cognitively, but we also feel and evaluate trust through our emotional senses (Lewis & Weigert, 1985). There is a direct relationship between the building of trust and the recognition and functional use of the emotional side of human behavior. McAllister (1995) suggested that the role of emotion is critical in explaining how and under what circumstances trust turns to distrust. Defensive routines (Argyris, 1990) are an example of an emotive response that can impede the trust development process or affect the preservation of existing trust levels. Recent descriptions of emotional intelligence (Goleman, 1995) provide a mechanism for exploring and understanding the emotional side of trust and legitimate emotional sensing as an input to affective decision-making. High EQ (emotional intelligence) is often associated with individuals who are predictably more trusting and organizations that have supportive infrastructures that recognize that behavior, human and organizational, have both cognitive and affective components. Recognition of the emotional aspect of trust and the trust-building process helps the understanding and efficiency of trust development.

Seventh, the context or situation in which a trust transaction occurs is an essential ingredient affecting the eventual judgment of the trust level. Context refers to the environmental conditions that exist during any trust transaction. An example of a contextual element in a workplace trust transaction might be the day of the week that an action occurs. Using the example of an employee who downloads software for personal use through a work computer, the context of this action — during working hours, outside of working hours, on a weekend or a workday — has a varying effect on the resultant trust level. Depending upon the context, interactions might take wholly different paths which will lead to differing outcomes. In our example, a weekend or after-hour breach of company policy might have a lesser effect on the ultimate trust level if the employee used work time or flaunted the policy when his co-workers were present. The difference in context might involve who knew about the protocol violation. Context is a challenging concept in which to develop a shared meaning or common understanding (McKnight, Cummings, & Chervany, 1996, 1998; McKnight & Chervany, 1996). Oftentimes, the contextual views of two individuals experiencing the exact same interaction are dramatically different. This can be further illustrated through a simple observational experiment.

Give a picture to two people and tell them to describe a story based on what they see. You are most likely going to get two significantly different stories based on the varying latent context perceived by the observers. Context also has an effect on the ethics of trust. Hosmer (1995) suggests that morally correct decisions and actions are based upon ethical principles of analysis which establish an expectation of ethically justifiable behavior. In critical human situations when life or property is at a high risk, the context affects the degree of acceptability of interactions.

Context is oftentimes overlooked, and in the case of building trust within organizations, context is frequently the critical difference between success and failure. Overt actions need to occur at the 'right time' or within the right situation. Attention to context and consciousness of situational difference are vital to assisting in the trust-building and trust-development processes.

Eighth, companies have an opportunity to guide their employees, with respect to using the Internet for non-work-related activities, by establishing and publishing codified acceptability procedures. The codification of protocols (policy) provides a written description of what is and is not acceptable within the workplace. These protocols enable the employee to review the corporate guidelines and make choices based on what management believes to be appropriate within their work environment. When codified procedures are not available, the employee reacts based on tacit expectations, personal character, referent group common practice, and individual ethical beliefs. The use of implicit guidelines places the burden of responsibility solely on the employee to behave appropriately when written policy is not available. Therefore, an organizational best practice is the creation of policy and procedures on the use of company equipment and in particular the Internet which are explicit and widely distributed.

Companies can reinforce the constructive use of the Internet through established guidelines. Policy development offers guidance to employees regarding what websites or type of activities are acceptable when undertaken using company equipment. However, the establishment and publication of policy is not enough. Companies need a mechanism for ensuring that organizational policy, referent group ethical standards, and special use procedures are known and understood by all employees. Company internal communication and monitoring systems can serve as reinforcements of the written protocols. Automated monitoring and control systems such as firewalls limit employee Web access from a corporate site. Additionally, corporate agents might supply employees with a list of restricted URLs or subject matter considered inappropriate. The key to avoiding unauthorized or inappropriate Web usage by company employees is to be preventive by proactively establishing the ethics of personal Web usage rather than dealing with violations as they occur.

Ninth, one mechanism for developing a culture that has great promise for organizations struggling to establish and maintain a climate of trust is the use of explicit modeling. Social identity theory suggests that people identify with those who are part of the same referent group (Hogg & Terry, 2000). The implication of social identity theory on trust is that individuals behave consistent with their observation of their referent group (Ashborth & Mael, 1989). In other words, employees in a work context will take their cues from their superiors, subordinates, and peers (Berscheid, 1985).

Generally, referent groups are aligned by values, belief systems, regional geography, ideology, interest, job function, and/or areas of expertise (Turner, 1982). Everyone has many referent groups to which they belong and align to a greater or lesser degree (Turner, 1985). For example, managers at strategic levels will consider other executives within their referent group; or information technology user groups within a geographical (local, regional) area might be groups of individuals with an interest, knowledge, and aptitude for using technology. Therefore, the individuals within the user group, even if from different geographic settings, would be part of the referent group. Members of any community which hold some things in common — neighborhood, interest group, professional association, club, department within a company, religious belief system, political party — are examples of referent groups. Referent groups are the single greatest source of modeling within organizations.

Organizations operationalize modeling in order to nurture and maintain a culture of trust beginning with managerial awareness. Understanding the nature of trust in the work environment and how trust manifests is foundational to establishing and using modeling as a mechanism for organizational trust development. Modeling requires conscious awareness, explicit intent to behave in a consistent manner (in this case trustworthy), and a dedication to use available opportunities to demonstrate and 'model' what constitutes trust behaviors within an organization. Challenges to perceptions of distrust must be confronted directly and immediately.

For example, managers may have repeated experiences with lower-level employees that are negative. This may produce judgments by managers that lower-level employees lack integrity; are less trustworthy than others within the company; or do, can, or will not live up to an expected standard. As such, the manager may have a predisposition to be less trusting of these individuals than for operational or managerial employees. This perception must be confronted if an atmosphere of trust is to be fostered. Unfortunately, if the belief that trust of an individual or group is not warranted, the result is often that a culture of distrust is unwittingly precipitated and reinforced. Therefore, organizations must deal with the presumption of trust, the perceptions of trust, the judgments about trust, and finally have a procedure/protocol to deal with trust anomalies, both perceived and actual.

RECOMMENDATIONS FOR MANAGERS TO REINFORCE TRUST IN THE WORKPLACE

Managers are constant role models. It is through role modeling that trust cultures are established and nurtured in organizations. Employees often look toward their managers to determine what constitutes appropriate behavior. Employees need to be able to observe their managers to see what is and is not acceptable behavior within an organizational setting. It is important that managers accept and model appropriate Web use behavior congruent with organizational policies. Unfortunately, it is often difficult for managers to determine when to trust, who to trust, and how much to trust (Kramer, 1995). The following are some recommended considerations for corporate leaders to employ in reinforcing trust in the workplace.

1. Establish a specific set of standards and policies for non-work-related Internet activities.
2. Conduct organizational orientations on using the Internet for non-work-related reasons.
3. Carefully integrate corporate policies into organizational handbooks.
4. Make the issue of interpersonal trust part of strategic planning and annual goal setting.
5. Ensure that human resource professionals within the company and other organizational leaders have a common understanding and agreement about the importance of trust, the nature of organizational trust, and an assessment of the climate of trust that exists.
6. Explicitly assess interpersonal trust annually.
7. Establish what constitutes the standards or ethics of individual behavior related to trust. Use of the Internet for non-work-related activities should be considered and standardized throughout the organization. In some organizations, a 100% open use policy will be acceptable. In others, if an employee uses technology during work hours for non-work related-activities, this is a form of unauthorized appropriation of company resources. Any time used for personal work in this context is time owed back to the company. In a few organizations, a 100% non-use policy might be required.
8. Ensure that written policy and procedures are developed and used for any breaches in interpersonal trust.
9. In every opportunity that arises, consistently model trustworthy behavior, because "employees do what managers do, before they do what managers say." Set a good example: everyone — managers and subordinates — needs to be good role models.
10. Hire trustworthy people, by making a preliminary trust assessment a condition of employment. Trustworthiness is as important a workplace competency as any technical or functional skill.
11. Talk about policies of acceptability (norms of behavior, expectations of management) informally.

12. Establish a culture of trust where trustworthy behavior is acknowledged, rewarded, and praised.
13. Finally, develop trust heroes, i.e., visible symbols within and outside of the company who are exemplars of trustworthy behavior in organizational settings.

Assuming that the use of the Internet for some non-work-related activity is acceptable within company guidelines, the individual must ultimately decide the appropriate level of involvement and use. The organization must temporarily risk trusting the individual employee to act appropriately within the guidelines. The organizational trust culture can be self-fulfilling if the organization: (1) places and praises trust in appropriate use; and (2) makes trust explicit, resulting in organizational members behaving and performing as expected and intended. In the final analysis, the degree of trust placed upon an employee is conferred by giving him Internet access for completion of his daily tasks.

There is an implicit expectation in some organizations that the employee will behave responsibly and that the individual will use the equipment in a trustworthy manner. The implication of these behaviors is that the employee does not place himself nor his employer in any ethical, moral, or legal situation based on non-trustworthy behavior. The dilemma is that much, if not all, of the information exchanged while at an office is the property of the company by common business practice. The risk of a proprietary violation is real, and recent public cases of trust violations that have resulted in protracted litigation do exist. The fallout from some of these highly publicized cases is an ever-increasing set of corporate security practices.

Another common organizational practice in the information age has been the retention of all e-mail that goes from or comes to a corporate site. If a trust violation occurs at a later time, this archival information has been or can be used against employees. Careful consideration must be given to all policies related to the ethics of personal Web use for each individual organization.

It is not easy to deal with trust in the workplace and even more difficult to operationalize the concept as part of business practice. However, if trust is to be considered as an important element of organizational culture, then nothing short of full attention to trust is required. Managers need to avail themselves of every possible means to promote trustworthiness and every opportunity to demonstrate trust, since trust in management is closely tied to productivity outcomes (Davis, Mayer, & Schoorman, 1995). Organizational leaders need to model trust in every practice, both internally and externally, and trust must be made part of the everyday 'dialogue' of the organization. Trust should be accepted as part of the corporate strategic goals, and workplace practices that deal with trust must be encouraged and discussed. Everyone must act as a role model for their referent groups in modeling trustworthy behavior in the workplace environment.

SUMMARY

Trust in the workplace is a phenomenon that is illusive and challenging to build and sustain. Trust can be viewed in a variety of different ways: as the outcome of transactions, as a mechanism that supports the quality of interactions, or as the perception, evaluation, or judgment that filters individual and organizational thoughts, emotions, policies, procedures, and practices. Trust is sometimes viewed solely as an interpersonal issue. Trust as an organizational issue has not been fully developed in the literature, primarily because of the complexity, abstract nature, and variance in meaning. However, when we examine individual employee practices within organizational or workplace settings, we are frequently confronted with the notion of trust as a latent variable that has a profound effect on product and process, and hence a business' bottom line. Particularly as addressed in this text, we are concerned with trust as an organizational issue in the use of technology within a work setting for personal, non-work-related activities. To some degree, this is a matter of policy, of ethics, of standards, and ultimately of organizationally defined acceptable practice. But, the use of the Web for personal activity or business is clearly related to the trust which exists between individuals, and between the organization and individual employees.

A case is made for the development of trust within organizations as a corporate goal through a specific set of activities or actions. Evidence exists that the costs of doing business can be greatly reduced when trust becomes an active and explicit organizational consideration. Trust can be developed as an element of the organizational culture, through building a climate of trust and supporting trustworthiness by organizational members. As such, trust, in this context, is definitively a human resource management issue.

We build trust through awareness, consciousness, and action. We can learn to understand that trust exists at various levels and that each level — calculative-based trust, knowledge-based trust, and identification-based trust — requires different techniques to establish, maintain, and enhance trust between individuals at the respective level. Companies can become oriented in such a way that trust is explicitly addressed, measured, and consciously made part of everyday operations.

A successful technique for the development and maintenance of trust within organizations is behavior modeling. Not simple role modeling alone, but rather the long-term modeling of trustworthy behavior by members of referent groups. Trustworthy behavior supports and encourages acceptable actions by organizational members. As effective role models, employees are capable of building an infrastructure or climate of trust through: (1) individual interaction processes; (2) conscious dialogue; and (3) consistent and predictable behavior.

Trust may reduce operating costs. Trust can be developed, nurtured, modeled, maintained, measured, modified, molded, and managed. The personal use of the Web in the workplace by employees is but a single example of an application of trust to the workplace. Through understanding and commitment, organizations and individuals can use trust to help manage acceptable employee non-work activities through the Internet and also generalize trust to other organizational and business processes, practices, and products.

REFERENCES

Argyris, C. (1964). *Integrating the Individual and the Organization*. New York: John Wiley & Sons.

Argyris, C. (1990). *Overcoming Organizational Defenses*. Boston, MA: Allyn & Bacon.

Argyris, C. & Schöen, D.A. (1978). *Organizational Learning*. Reading, MA: Addison-Wesley.

Arrow, K. (1974). *The Limits of Organization*. New York: Norton.

Ashborth, B.E. & Mael, F. (1989). Social identity theory and the organization. *Academy of Management Review*, 14(1), 20-39.

Barber, B. (1983). *The Logic and Limits of Trust*. New Brunswick, NJ: Rutgers University Press.

Berscheid, E. (1985). Interpersonal attraction. In Lindzey, G. & Aronson, E. (Eds.), *Handbook of Social Psychology* (Vol. II, pp. 413-484). New York: Random House.

Bhattacharya, R., Devinney, T.M., & Pillutla, M.M. (1998). A formal model of trust based on outcomes. *Academy of Management Review,* 23(3), 459-472.

Blau, P.M. (1968). Interaction: Social exchange. *International Encyclopedia of the Social Sciences*. New York: The Free Press and Macmillan.

Boon, S.D. & Holmes, J.G. (1991). The dynamics of interpersonal trust resolving uncertainty in the face of risk. In Hinde, R.A. & Groebel, J. (Eds.), *Cooperation and Prosocial Behavior* (pp. 190-211). Cambridge, UK: Cambridge University Press.

Bromiley, P. & Cummings, L.L. (1995). Transaction costs in organizations with trust. In Lewicki, R.J., Bies, R.J., & Sheppard, B.H. (Eds.), *Research on Negotiation in Organizations* (Vol. 5, pp. 219-247). Greenwich, CT: JAI Press.

Butler, J.K. (1991). Toward Understanding and measuring conditions of trust: Evolution of a condition of trust inventory. *Journal of Management*, 17(3), 643-663.

Butler, J.K. & Cantrell, R.S. (1984). A behavioral decision theory approach to modeling dyadic trust in superiors and subordinates. *Psychological Reports*, 55, 19-28.

Coleman, J.S. (1990). *Foundations of Social Theory.* Cambridge, MA: Harvard University Press.

Cook, J. & Wall, T. (1980). New work attitude measures of trust, organizational commitment and personal need non-fulfillment. *Journal of Occupational Psychology,* 53(1), 39-52.

Cummings, L.L. & Bromiley, P. (1996). The Organizational Trust Inventory (OTI): Development and validation. In Kramer, R.M. & Tyler, T.R. (Eds.), *Trust in Organizations: Frontiers of Theory and Research* (pp. 302-330). Thousand Oaks, CA: Sage Publications.

Dasgupta, P. (1988). Trust as a commodity. In Gambetta, D. (Ed.), *Trust: Making and Breaking Cooperative Relations* (pp. 49-72). New York: Basil Blackwell Ltd.

Davis, J.H., Mayer, R.C., & Schoorman, F.D. (1995). The trusted general manager and firm performance: Empirical evidence of a strategic advantage. Paper presented at the *International Strategic Management Society Meetings*, October, Mexico City, Mexico.

Deutsch, M. (1958). Trust and suspicion. *Conflict Resolution*, II(4), 265-279.

Deutsch, M. (1960). The effect of motivational orientation upon trust and suspicion. *Human Relations*, 13, 122-139.

Deutsch, M. (1973). *The Resolution of Conflict: Constructive and Destructive Processes.* New Haven, CT: Yale University Press.

Fukuyama, F. (1995). *Trust: The Social Virtues and the Creation of Prosperity*. New York: The Free Press.

Gabarro, J. (1987). *The Dynamics of Taking Charge*. Boston, MA: Harvard Business School Press.

Gambetta, D. (1988). Can we trust trust? In Gambetta, D. (Ed.), *Trust: Making and Breaking Cooperative Relationships* (pp. 213-237). New York: Basil Blackwell.

Giffin, K. (1967). The contribution of studies of source credibility to a theory of interpersonal trust in the communication process. *Psychological Bulletin*, 68(2), 104-120.

Goleman, D. (1995). *Emotional Intelligence: Why It Can Matter More Than IQ*. New York: Bantam Books.

Good, D. (1988). Individuals, interpersonal relations, and trust. In Gambetta, D. (Ed.), *Trust: Making and Breaking Cooperative Relations* (pp. 31-48). New York: Basil Blackwell.

Gulati, R. (1995). Does familiarity breed trust? The implications of repeated ties for contractual choice in alliances. *Academy of Management Journal*, 38(1), 85-112.

Hogg, M.A. & Terry, D.J. (2000). Social identity and self-categorization processes in organizational contexts. *Academy of Management Review*, 25(1), 121-140.

Holmes, J.G. (1991). Trust and the appraisal process in close relationships. In Jones, W.H. & Perlman, D. (Eds.), *Advances in Personal Relationships* (Vol. 2, pp. 57-104). London: Jessica Kingsley.

Hosmer, L.T. (1995). Trust: The connecting link between organizational theory and philosophical Ethics. *Academy of Management Review*, 20(2), 379-403.

Johnson-George, C. & Swap, W.C. (1982). Measurement of specific interpersonal trust: Construction and validation of a scale to access trust in a specific other. *Journal of Personality and Social Psychology,* 43(6), 1306-1317.

Kramer, R.M. (1995). Power, paranoia and distrust in organizations: A distorted view from the top. In Lewicki, R.J., Bies, R.J., & Sheppard, B.H. (Eds.), *Research on Negotiation in Organizations* (Vol. 5, pp. 119-154). Greenwich, CT: JAI Press.

Larzelere, R.E. & Huston, T.L. (1980). The dyadic trust scale: Toward understanding interpersonal trust in close relationships. *Journal of Marriage and the Family*, 42(3), 595-604.

Lewicki, R.J. & Bunker, B.B. (1995). Trust in relationships: A model of development and decline. In Bunker, B.B. & Rubin, J.Z. (Eds.), *Conflict, Cooperation and Justice* (pp. 133-173). San Francisco, CA: Jossey-Bass.

Lewicki, R.J. & Bunker, B.B. (1996). Developing and maintaining trust in work relationships. In Kramer, R.M. & Tyler, T.R. (Eds.), *Trust in Organizations: Frontiers of Theory and Research* (pp. 114-139). Thousand Oaks, CA: Sage Publications.

Lewis, J.D. & Weigert, A. (1985). Trust as a social reality. *Social Forces*, 63(4), 967-985.

Lippert, S.K. (2001). *An Exploratory Study into the Relevance of Trust in the Context of Information Systems Technology.* Doctoral Dissertation, The George Washington University, Washington, DC.

Lippert, S.K. (2002). Contributing to a unified model of technology trust: Understanding trust in information systems technology. Presented at the *2002 Academy of Business and Information Technology Meeting*, (May 2-4), Pittsburgh, PA.

Luhmann, N. (1979). *Trust and Power*. Chichester, UK: John Wiley & Sons.

Mayer, R.C., Davis, J.H., & Schoorman, F.D. (1995). An integrative model of organizational trust. *Academy of Management Review*, 20(3), 709-734.

McAllister, D.J. (1995). Affect- and cognition-based trust as foundations for interpersonal cooperation in organizations. *Academy of Management Journal*, 38(1), 24-59.

McGreger, D. (1967). *The Professional Manager*. New York: McGraw Hill.

McKnight, D.H. & Chervany, N.L. (1996). The meanings of trust. *MISRC Working Paper Series*, University of Minnesota. Working Paper 96-04. (www.misrc.umn.edu/wpaper/wp96-04.htm).

McKnight, D.H., Cummings, L.L., & Chervany, N.L. (1996). Trust formation in new organizational relationships. *MISRC Working Paper Series*, University of Minnesota. Working Paper 96-01. (www.misrc.umn.edu/wpaper/wp96-01.htm).

McKnight, D.H., Cummings, L.L., & Chervany, N.L. (1998). Initial trust formation in new organizational relationship. *Academy of Management Review*, 23(3), 473-490.

Mellinger, G.D. (1956). Interpersonal trust as a factor in communication. *Journal of Abnormal Social Psychology*, 52, 304-309.

Mishra, A.K. (1996). Organizational responses to crisis: The centrality of trust. In Kramer, R.M. & Tyler, T.R. (Eds.), *Trust in Organizations: Frontiers of Theory and Research* (pp. 261-287). Thousand Oaks, CA: Sage Publications.

Rempel, J.K. & Holmes, J.G. (1986). How do I trust thee? *Psychology Today,* (February), 28-34.

Rempel, J.K., Holmes, J.G., & Zanna, M.P. (1985). Trust in close relationships. *Journal of Personality and Social Psychology*, 49(1), 95-112.

Robinson, S.L. (1996). Trust and breach of the psychological contract. *Administrative Sciences Quarterly*, 41, 574-599.

Rotter, J.B. (1967). A new scale for the measurement of interpersonal trust. *Journal of Personality,* 35(4), 651-665.

Rotter, J.B. (1971). Generalized expectancies for interpersonal trust. *American Psychologist,* 26(5), 443-452.

Schlenker, B.R., Helm, B., & Tedeschi, J.T. (1973). The effects of personality and situational variables on behavior trust. *Journal of Personality and Social Psychology*, 25, 419-427.

Schneider, B. (1990). *Organizational Climate and Culture.* San Francisco, CA: Jossey-Bass.

Schoorman, F.D, Mayer, R.C., & Davis, J.H. (1996). Empowerment in veterinary clinics: The role of trust in delegation. Paper Presented at the *Annual Meeting of Society for Industrial and Organizational Psychology*, San Diego, CA.

Shapiro, D., Sheppard, B.H., & Cheraskin, L. (1992). Business on a handshake. *Negotiation Journal*, 8(4), 365-377.

Sheppard, B.H. & Tuchinsky, M. (1996). Micro-OB and the network organization. In Kramer, R.M. & Tyler, T.R. (Eds.), *Trust in Organizations: Frontiers of Theory and Research* (pp. 140-165). Thousand Oaks, CA: Sage Publications.

Sitkin, S.B. & Stickel, D. (1995). The road to Hell: The dynamics of distrust in an era of quality. In Kramer, R.M. & Tyler, T.R. (Eds.), *Trust in Organizations: Frontiers of Theory and Research* (pp. 196-215). Thousand Oaks, CA: Sage Publications.

Turner, J.C. (1982). Toward a cognitive redefinition of the social group. In Tajfel, H. (Ed.), *Social Identity and Intergroup Relations* (pp. 1-40). Cambridge, UK: Cambridge University Press.

Turner, J.C. (1985). Social categorization and the self-concept: A social cognitive theory of group behavior. In Lawler, E.L. (Ed.), *Advances in Group Processes* (Vol. 2, pp. 77-122). JAI Press.

Tyler, T.R. & Degoey, P. (1995). Trust in organizational authorities: The influence of motive attributions on willingness to accept decisions. In Kramer, R.M. & Tyler, T.R. (Eds.), *Trust in Organizations: Frontiers of Theory and Research* (pp. 331-356). Thousand Oaks, CA: Sage Publications.

Williamson, W.E. (1975). *Markets and Hierarchies: Analysis and Antitrust Implications*. New York: The Free Press.

Zaheer, A., McEvily, B., & Perrone, V. (1998). Does trust matter? Exploring the effects of interorganizational and interpersonal trust on performance. *Organization Science*, 9(2), 141-159.

Zand, D.E. (1972). Trust and managerial problem solving. *Administrative Science Quarterly,* 17(2), 229-239.

Zucker, L.G. (1986). Production of trust: Institutional sources of economic structure, 1840-1920. In Staw, B.M. & Cummings, L.L. (Eds.), *Research in Organizational Behavior* (Vol. 8, pp. 53-111). Greenwich, CT: JAI Press.

Chapter VI

A Deterrence Theory Perspective on Personal Web Usage

Dinesh A. Mirchandani
University of Missouri - St. Louis, USA

ABSTRACT

Personal Web usage (PWU) in the workplace is a matter of considerable concern to organizations today. However, human resources managers are not fully aware of the range of actions they can take to reduce PWU. This chapter examines general deterrence theory in the context of PWU and identifies actions that managers can take to reduce PWU. It uses a two-stage research methodology consisting of: (1) interviews with managers to gather qualitative data, and (2) a field survey of end users to gather quantitative data on PWU. It finds support for the deterrence theory and recommends that managers use a sequential implementation of four deterrence stages to contain PWU. These are: (1) framing an 'Internet use policy,' (2) implementing measures to prevent PWU such as restricting Internet access only to certain employees, (3) implementing appropriate content management software to detect PWU, and (4) standardizing policies to remedy non-acceptable Web usage. These four deterrence stages can protect an organization from the harmful effects of PWU of its employees.

INTRODUCTION

Personal Web usage (PWU) in the workplace is defined as "any voluntary act of employees using their companies' Internet access during office hours to surf non-work-related websites for non-work purposes" (Lim, Teo, & Looh, 2002). The occurrence of PWU may be viewed as a kind of systems risk, i.e., the likelihood that a firm's information systems (IS) are insufficiently protected against certain kinds of damage or loss. As with systems risk, managers are generally unaware of the full range of actions they can take to reduce PWU (Straub & Welke, 1998; Lim, Teo, & Looh, 2002). The general deterrence theory, drawn from the field of criminology, suggests that sanctions and disincentive measures can reduce systems abuse by making potential abusers aware that their unethical behavior will be detrimental to their own good (Pearson & Weiner, 1985).

According to this theory, strategies that can be adopted to reduce systems risk fall into four distinct, sequential activities, i.e., (1) deterrence, (2) prevention, (3) detection, and (4) remedies. *Deterrent* measures include policies and guidelines for proper system use. They tend to be passive in that they have no inherent provision for enforcement. They depend wholly on the willingness of system users to comply. *Preventive* measures include for example locks on computer room doors and password access controls. They are active measures with inherent capabilities to enforce policy and ward off illegitimate use. If deterrent and preventive measures are unsuccessful in containing abuse, then *detection* measures can be deployed. These include proactive security responses such as suspicious activity reports, system audits and virus scanning reports, or reactive responses such as detective work after a documented breach in security. These measures gather evidence of abuse and identify perpetrators. Finally, *remedies* are measures that can correct the harmful effect of an abusive act and punish the perpetrators. Internal actions include warnings, reprimands, and termination of employment. Legal actions include criminal and civil suits (Straub & Welke, 1998). Thus a company can start by deploying deterrent measures. If these are not successful in warding off abusers, the company can then use preventive, detective, and finally remedial measures, which in that order proceed from being milder to stronger. There is, however, limited evidence available in practice to prove the effectiveness of these four techniques despite their strong theoretical basis. This research thus seeks to empirically support the general deterrence theory in the context of PWU. Its major contribution lies in the implications it provides for practicing managers and researchers.

BACKGROUND

Information systems risk has continued to be a problem in the decades of computerization primarily because managers have either ignored the issue or not versed themselves on its nature (Forcht, 1994; Loch, Carr, & Warkentin, 1992). PWU in organizations opens the door to mission-critical information systems becoming unavailable for basic transactions due to network congestion or virus attacks (Straub & Welke, 1998), in addition to lost productivity and employee discontent (Verespej, 2000; Marsan, 2000), and therefore is clearly a systems risk. Yet, no clear strategy has been developed to cope with this risk, suggesting the need in organizations to define a new role for a Chief Privacy and Integrity Officer (CPIO) responsible for formulating the Internet use policy, implementing defenses against the abuse, and securing adequate liability insurance (Sipior & Ward, 2002).

However, defenses against PWU can only be effective if top management is firm in its support for these controls, provides adequate training, promotes user involvement to create acceptance, and develops organizational citizenship behavior (OCB) in the employees (McGowan & Klammer, 1997; Holmes et al., 2002). Important dimensions of OCB include behaviors reflecting concern for the firm (civic virtue) and behaviors going beyond simply obeying rules (conscientiousness) (Organ, 1988). Promoting these kinds of behaviors with organizational reward practices can also reduce systems risk (Schnake & Dumler, 1997). If managers on the other hand choose to enforce the defenses through monitoring and surveillance of employees, they run the risk of eroding mutual trust between leaders and employees who may then engage in opportunistic behavior. Unambiguous policies and procedures that are perceived to be fair and equitably administered stand a better chance of acceptance within the organization (Holmes et al., 2002).

However, IS managers' knowledge of technical and managerial controls that can cope with systems risk is fragmentary and incomplete (Straub & Welke, 1998). Consciousness of risk-lowering actions can lead to effective planning and implementation of defenses against PWU. Managers' consciousness about security risk can be heightened by a firm grasp of the level of systems risk that the industry as a whole is exposed to, a basic understanding of actions that can be taken to cope with risk, and by being well informed of the local incidence of abuse and the susceptibility to damage within the organization (Goodhue & Straub, 1991). This study seeks to raise that consciousness by identifying for managers the actions they can take to cope with PWU. It uses the general deterrence theory to put perspective on these actions.

THEORY

The general deterrence theory has been widely used to study the behavior of criminal and antisocial elements in both economics and criminology (Becker, 1968; Pearson & Weiner, 1985). The theory follows the notion that human behavior results from the pursuit of pleasure and the avoidance of pain; and to deter potential criminals from committing offences, it is necessary to impose sanctions that increase the costs and/or reduce the benefits associated with doing so (Becker, 1968).

Hence, the modern economic theory of crime is based on the assumption that rational individuals act to maximize their utility given the possibility of assigning time or resources to different activities. As such, there is no set group of individuals who are criminals. Rather, individuals move in and out of illegal activities as their opportunities change. A decision whether to undertake criminal activity is made taking into account the benefits and costs of alternative forms of action. An increase in the probability and/or severity of punishment (representing costs of criminal behavior) will reduce the potential criminal's participation in illegitimate activities (Bodman & Maultby, 1997).

In addition to postulating rational behavior on the part of potential criminals, the theory also incorporates the notion of rational, maximizing behavior on the part of the victims of the crime. Rational potential victims (i.e., organizations) attempt to minimize the costs of crime with the use of police resources, and the higher the crime rate, the greater will be the quantity of resources allocated to catch and punish perpetrators (Loftin & McDowell, 1982).

In reducing systems risk, researchers have proposed four distinct, sequential activities which are grounded in the principles of the deterrence theory (Forcht, 1994; Straub & Welke, 1998). These are: (1) deterrence, (2) prevention, (3) detection, and (4) remedy. According to Straub and Welke (1998), some system abuse is allayed by deterrent techniques such as policies and guidelines for proper system use, and by reminders to users to change their passwords. Deterrent measures tend to be passive and do not have an inherent provision for enforcement. They depend entirely on the willingness of system users to comply. Security awareness programs are an example of deterrent measures because these sessions: convey to users and managers knowledge about risks in the organizational environment; emphasize actions taken by the firm, including policies and sanctions for violations; and reveal threats to local systems and their vulnerability to attack. A major reason for initiating this training is to convince potential abusers that the company is serious about protecting its systems and will punish abusers.

Deterrent measures from the perspective of PWU would include policy statements, educational seminars, and other passive methods of making employees aware of the severity of the crime and its punishment.

When potential abusers choose to ignore deterrents, preventive measures such as locks on computer room doors and password access controls can defend systems from abusers. Preventives thus are active measures that have the capability to enforce policy and ward of illegitimate use. The use of filtering software can restrict access to certain websites and actively prevent PWU (Roberts, 1999).

If system abusers are not warded off by deterrent and preventive measures, the organization needs to be able to detect misuse. Proactive security responses such as suspicious activity reports, system audits, and virus scanning are examples of these. Reactive responses include detective work after a documented breach in security. The primary objective of this security response is to gather evidence of misuse and to identify perpetrators. By monitoring the websites visited as well as the emails of suspicious employees, the organization can build an effective case of abuse against them.

Lastly, the security program should be able to correct the harmful effects of abuse and punish offenders. Internal actions can include warnings, reprimands, and termination of employment, while legal actions can include criminal and civil suits.

From the perspective of the general deterrence theory, these four kinds of defense can reduce PWU. The potential abusers become convinced of the certainty and severity of punishment for committing certain acts when the effectiveness of the system security is obvious or communicated to them.

Because there is insufficient evidence to prove the effectiveness of these four techniques despite their theoretical basis (Straub, 1990), the current study sought to empirically test the following hypothesis.

H_0: *There will be no difference in the perceived effectiveness of deterrent, preventive, detective, and remedial techniques in reducing PWU.*

H_a: *Deterrent techniques will be perceived to be less effective than preventive, detective, and remedial techniques in reducing PWU.*

Deterrent techniques being passive in nature without any provision for enforcement will be less likely to ward off PWU than active techniques like preventives, detectives, or remedies. Quite clearly, potential system abusers will refrain from such actions if the organization deploys active measures that increase the likelihood of their being identified and punished.

METHODOLOGY

A two-stage methodology was adopted for this study that consisted of: (1) interviews with managers to gather qualitative data, and (2) a field survey of end users to gather quantitative data. Structured interviews were first conducted with managers of 66 companies that provided their employees Internet access in the workplace. Forty-six of these companies were in the service sector, whereas the remaining were in the manufacturing sector. These companies ranged in size from small (<500 employees, n = 38, m = 113 employees), to medium (between 500 and 1,000 employees, n = 16, m = 871 employees), and large (>1,000 employees, n = 12, m = 12,517 employees). Of the managers interviewed, eight were IS managers and the remaining were non-IS managers. These managers were asked to describe the measures their companies were taking to reduce PWU at work by the employees. A total of 18 measures were identified by the managers as actions their companies took to reduce PWU.

These 18 measures were used to develop a survey instrument for use in the second stage of the study, in which respondents could choose from 1 to 7 (1 being 'not effective' and 7 being 'very effective'). Likert-type scales their opinion on how effective each of the 18 measures would be in reducing PWU. Seven end

Table 1. Measures that Companies are Taking to Reduce PWU

Measures that Companies are Taking to Reduce PWU	Percentage of Companies Using this Measure
Deterrent Measures	
To have a written company manual/policy sheet/employee handbook/memorandum stating that the Internet at work is to be used for work-related purposes only	42.2
To have employees who access Internet-enabled computers at work to log their name, time in, time out, and the reason for using the Internet	6.8
To arrange seminars, staff meetings, and show videotapes to educate new and old employees about Internet abuse	4.7
To have employees with Internet access at work fill out weekly log sheets describing their Internet usage	1.6
Preventive Measures	
To limit Internet access to only certain employees upon their supervisors' consent	25.0
To block access to non-work-related and offensive websites by using Internet filters	23.4
To have employees sign forms stating that they will abstain from visiting offensive websites while at work	21.4
To have employees agree to accept the company's 'Internet Use Policy' when logging into their computers	18.7
To allow but limit personal Internet usage to employees in their free time, or after work hours, or in emergencies	17.2
Detective Measures	
To monitor with special software all the websites visited by employees	26.6
To monitor with special software all the emails of employees	20.8
To monitor electronic files downloaded on the computers of employees to identify if they are non-work-related	16.7
To monitor with special software what every computer in the company is being used for at a particular time	15.1
To use an 'Internet cop' to police the workplace for Internet abuse	5.7
To watch on cameras all employees using computers	0.5
Remedial Measures	
To have managers reprimand employees who abuse the Internet at work	29.2
To take away Internet privileges of employees who abuse the Internet at work	24.0
To terminate employees who abuse the Internet at work	22.4

users affiliated with an organization that provides them Internet access at work volunteered to participate in a pilot of the survey instrument. These subjects were provided definitions of deterrent, preventive, detective, and remedial measures, and under the facilitation of the author discussed the nature of the 18 measures, and by a group consensus classified each into one of the four categories. The measures and their classification is shown in Table 1.

In the second stage of the study, three survey instruments were personally administered by each student (total number of students = 72) enrolled in an undergraduate management information systems class of a mid-western university in the U.S. to three white-collar office workers that the student was acquainted with. To ensure unique responses, each respondent was also required to list his/her name and contact telephone number at the end of the survey. Completed surveys came from 192 subjects for a response rate of 89%. The subjects rated the perceived effectiveness of each of the 18 measures in reducing PWU in their companies.

Table 2. Perceived Effectiveness of Measures to Reduce PWU

Perceived Effectiveness of Measures to Reduce PWU	Mean	Std. Dev.
To block access to non-work-related and offensive websites by using Internet filters	5.24	1.72
To terminate employees who abuse the Internet at work	5.11	1.94
To take away Internet privileges of employees who abuse the Internet at work	4.87	1.64
To monitor with special software all the websites visited by employees	4.85	1.53
To monitor with special software what every computer in the company is being used for at a particular time	4.61	1.82
To monitor electronic files downloaded on the computers of employees to identify if they are non-work-related	4.60	1.71
To have managers reprimand employees who abuse the Internet at work	4.57	1.58
To monitor with special software all the emails of employees	4.51	1.74
To limit Internet access to only certain employees upon their supervisors' consent	4.30	1.86
To allow but limit personal Internet usage to employees in their free time, or after work hours, or in emergencies	4.14	1.82
To have employees who access Internet-enabled computers at work to log their name, time in, time out, and the reason for using the Internet	3.92	1.70
To watch on cameras all employees using computers	3.89	2.16
To use an 'Internet cop' to police the workplace for Internet abuse	3.83	1.89
To have a written company manual/policy sheet/employee handbook/memorandum stating that the Internet at work is to be used for work-related purposes only	3.70	1.73
To have employees sign forms stating that they will abstain from visiting offensive websites while at work	3.60	1.69
To have employees agree to accept the company's 'Internet Use Policy' when logging into their computers	3.55	1.65
To arrange seminars, staff meetings, and show videotapes to educate new and old employees about Internet abuse	3.09	1.55
To have employees with Internet access at work fill out weekly log sheets describing their Internet usage	2.91	1.58

RESULTS

As the general deterrence theory predicts, measures that were preventive, detective, or remedial in nature were rated to be the most effective. Measures that tended to be deterrents were rated to be the least effective. The 18 measures ranked according to their perceived effectiveness are shown in Table 2.

The measure perceived to be most effective in reducing PWU was "to block access to non-work-related and offensive websites by using Internet

Table 3. Factor Analysis

Factor 1: Explicit Prevention and Detection (∝ =.87)	1	2	3	4
To monitor with special software all the websites visited by employees	.831	.151	-.04	.143
To monitor with special software what every computer in the company is being used for at a particular time	.814	.111	.137	.04
To monitor electronic files downloaded on the computers of employees to identify if they are non-work-related	.784	.122	.229	.130
To monitor with special software all the emails of employees	.778	.164	.206	.113
To block access to non-work-related and offensive websites by using Internet filters	.692	.130	.09	.194
Factor 2: Coerced Prevention and Detection (∝ = .80)				
To watch on cameras all employees using computers	.034	.808	.036	.195
To have employees with Internet access at work fill out weekly log sheets describing their Internet usage	.099	.792	.171	.092
To use an 'Internet cop' to police the workplace for Internet abuse	.157	.759	.147	.010
To have employees who access Internet-enabled computers at work to log their name, time in, time out, and the reason for using the Internet	.027	.749	.043	.325
Factor 3: Deterrence (∝ = .75)				
To have employees agree to accept the company's 'Internet Use Policy' when logging into their computers	.105	.08	.781	.253
To have a written company manual/policy sheet/employee handbook/memorandum stating that the Internet at work is to be used for work-related purposes only	.140	.03	.767	.118
To arrange seminars, staff meetings, and show videotapes to educate new and old employees about Internet abuse	.133	.126	.636	.252
To have employees sign forms stating that they will abstain from visiting offensive websites while at work	.140	.367	.613	-.07

Table 3. Factor Analysis (continued)

Factor 4: Remedies (∝ = .67)				
To take away Internet privileges of employees who abuse the Internet at work	.260	.09	.123	.792
To have managers reprimand employees who abuse the Internet at work	.155	.487	.117	.646
To allow but limit personal Internet usage to employees in their free time, or after work hours, or in emergencies	.109	.06	.321	.617
Eigen value	12.21	2.63	1.36	1.31
% of Total Variance Explained	35.48	12.16	9.83	6.89
Cumulative Variance Explained	35.48	47.64	57.47	64.36

filters," while the measure perceived to be least effective was "to have employees with Internet access at work fill out weekly log sheets describing their Internet usage." Active measures like preventives, detectives, and remedies were in general rated to be more effective than the passive deterrent measures. Thus, support was found for the alternative hypothesis that deterrent measures would be perceived to be less effective than the others.

Since factor analysis serves as a useful means of reducing a large number of items to a more manageable number and can make key themes visible (Nunnally, 1978), an exploratory factor analysis with varimax rotation was conducted on the 18 measures. The factor analysis revealed four factors and explained 64.36% of the variance in the data. Table 3 shows the factor loadings. Items with loadings greater than .5 were retained. Thus, of the 18 measures, two were dropped because of inconclusive loadings. These were "terminate employees who abuse the Internet at work" and "limit Internet access to only certain employees upon their supervisors' consent."

In performing the factor analysis, an iterative approach was followed of dropping inconclusive loadings as described by Sethi and King (1994). To remove any kind of subjectivity from the analysis, any measure that failed to load significantly on a factor was not retained. Cronbach's alpha reliability coefficients are shown in the table. The alpha values are greater than 0.6 indicating that the items in each factor do belong together. Table 3 also contains labels that the author applied to the factors. Deterrent and remedial measures showed up clearly as separate factors. However, preventive and detective measures did not appear to be distinct. In general though, the factors tied in well with the general deterrence theory and fairly close to the classification identified in Table 1. One factor that the author labeled as explicit prevention and

Table 4. Mean Perceived Effectiveness of Each Factor

Factor	Mean Perceived Effectiveness
1: Explicit prevention and detection	4.78
2: Remedies	4.53
3: Coerced prevention and detection	3.63
4: Deterrence	3.49

detection focused on the prevention and detection of PWU by monitoring the activities of employees. Another factor, labeled coerced prevention and detection, included measures that appeared intimidating with an intent to coerce employees into not abusing the Internet. A third factor called deterrence included four measures that appear passive and require the employee's cooperation. The last factor, named remedies, included two measures to punish offenders as well as one of cooperation with employees to allay the problem.

Table 4 shows the mean perceived effectiveness of each of the four factors. Once again, in support of the alternative hypothesis, the deterrents factor showed the lowest mean effectiveness compared to the other factors.

DISCUSSION AND IMPLICATIONS

This research helps identify for a practicing human resources manager the measures that may be the most effective in reducing PWU in the workplace. It also provides a perspective of what other companies and managers are doing to cope with this problem. An interesting revelation of the research is the wide breadth of measures that companies are using to reduce PWU. Some of these measures may even seem to infringe upon the rights of employees (such as monitoring their screens and keystrokes), and raise questions about corporate privacy policies. Should companies have to act as 'Big Brothers' to their employees? Or could other means be used to reduce PWU? For instance, several companies allow but limit personal Internet usage to employees in their free time, or after work hours, or in emergencies. These companies probably choose this informal approach towards personal Web usage to foster organizational citizenship among their employees. A 'kinder, gentler' company could in fact make its employees loyal and more productive.

On the other hand, knowledge that a particular employee is spending an inadvertently large amount of time on the Internet may suggest to the manager a deeper organizational problem, i.e., the employee is not being challenged enough by work assignments. Gifted employees with idle time on their hands are a wasted resource for the company. Thus detective measures, though intrusive upon the employees, could lead to positive outcomes for the company and the employees in terms of appropriate work assignments. Thus extrinsic behaviors of employees that appear to be non-productive may merely be symptoms of deeper, underlying problems. However, on the flip side of the coin, the addictive influence of the Internet on people has been well documented (Stanton, 2002). If intrinsically an employee becomes dependent on the Internet, which is a case not unlike substance abuse (Young, 1998), then detecting the harmful behavior early on would again be beneficial to the company as well as the employee.

The research shows that remedial measures such as reprimanding employees or curtailing their Internet usage are viewed among the most effective measures and are also widely used. Clearly then, punishment is an effective control mechanism. However, it would be interesting to see if compensation (monetary or otherwise) mechanisms are just as effective (Schnake & Dumler, 1997). None of the companies surveyed in this study used an employee compensation scheme to reduce PWU. However, future research should examine the possibility of using it.

An interesting point to note though is that the most widely used measure of having a written company policy barring PWU (utilized by nearly 42% of the respondent companies) is also considered to be one of the least effective. This clearly suggests that companies should give teeth to their policy statements for them to have any impact.

The blurred line between preventive and detective measures as seen in the factor analysis could be attributed to several causes. Perhaps the wording of some of the measures was confusing to the respondents. On the other hand, perhaps the content management software available to organizations today to curtail PWU is perceived as being both preventive as well as detective. In any case, future research should examine if indeed there is no distinction between the two, and possible reasons for this non-distinction.

In conclusion, it helps to visit some practical considerations. Most companies already have the means to track the PWU of their employees but choose not to do so simply because of the amount of effort involved in examining an Internet usage log, which can amount in effect to a full-time job for an IS staff member. Unless the average cost of PWU to a company in terms of lost productivity is estimated to be more than the annual salary of one employee, it may not be justifiable to add a new person to the IS department simply for containing PWU. If after this analysis, a human resources manager does decide to take on and contain PWU, the manager can do so by implementing four sequential deterrence stages as the theory suggests:

- First, an 'Internet use policy' needs to be framed to identify what constitutes acceptable and non-acceptable Web use, and this needs to be communicated to all employees.
- Second, the manager needs to institute preventive measures such as restricting Internet access only to employees who need it for their work.
- Third, an appropriate content management software needs to be identified and deployed to detect non-acceptable usage.
- Finally, policies need to be instituted to remedy non-acceptable Web usage.

These four sequential stages, when implemented, can provide an organization the best defense against PWU of employees.

REFERENCES

Becker, G. (1968). Crime and punishment: An economic approach. *Journal of Political Economy*, 76, 169-217.

Bodman, P. & Maultby, C. (1997). Crime, punishment and deterrence in Australia: A further empirical investigation. *International Journal of Social Economics,* 24(7/8/9), 884-898.

Forcht, K. (1994). *Computer Security Management.* Danvers, MA: Boyd and Fraser.

Goodhue, D. & Straub, D. (1991). Security concerns of system users: A study of the perceptions of the adequacy of security measures. *Information & Management*, 20(1), 13-27.

Holmes, S., Langford, M., Welch, O., & Welch, S. (2002). Associations between internal controls and organizational citizenship behavior. *Journal of Managerial Issues*, 14(1), 85-99.

Lim, V., Teo, T., & Looh, G. (2002). How do I loaf here? Let me count the ways. *Communications of the ACM*, 45(1), 66-70.

Loch, K., Carr, H., & Warkentin, M. (1992). Threats to information systems: Today is reality, yesterday is understanding. *MIS Quarterly*, 17(2), 173-186.

Loftin, C. & McDowell, D. (1982). The police, crime and economic theory. *American Sociological Review*, 47, 393-401.

McGowan, A. & Klammer, T. (1997). Satisfaction with activity-based cost management implementation. *Journal of Management Accounting Research*, 9, 217-238.

Merlino, L. (2000). Employers laid back over Internet abuse. *Upside*, 12(5), 46.

Nunnally, J.C. (1978). *Psychometric Research.* New York: McGraw-Hill.

Organ, D. (1988). *Organizational Citizenship Behavior.* Lexington, MA: Lexington Books.

Pearson, F.S. & Weiner, N.A. (1985). Toward an integration of criminological theories. *Journal of Crime and Criminology*, (Winter), 116-150.

Roberts, W. (1999). Filtering software blocks employees' Web abuses. *HRMagazine*, 44(9), 114-120.

Schnake, M. & Dumler, M. (1997). Organizational citizenship behavior: The impact of rewards and reward practices. *Journal of Managerial Issues*, 9(2), 216-229.

Sethi, V. & King, W. (1994). Development of measures to assess the extent to which an information technology application provides competitive advantage. *Management Science*, 40(12), 1601-1627.

Sipior, J. & Ward, B. (2002). A strategic response to the broad spectrum of Internet abuse. *Information Systems Management*, 19(4), 71-79.

Stanton, J. (n.d). Web addict or happy employee? Company profile of the frequent Internet user. *Communications of the ACM*, 45(1), 55-59.

Straub, D. (1990). Effective IS security: An empirical study. *Information Systems Research*, 1(3), 255-276.

Straub, D.W. & Welke, R.J. (1998). Coping with systems risk: Planning models for management decision making. *MIS Quarterly*, (December), 441-469.

Verespej, M.A. (2000). Internet surfing. *Industry Week*, 249(3), 58-64.

Young, K. (1998). Internet addiction: The emergence of a new clinical disorder. *CyberPsychology and Behavior*, 1(3), 237-244.

Chapter VII

Unsolicited Web Intrusions: Protecting Employers and Employees

Paulette S. Alexander
University of North Alabama, USA

ABSTRACT

Many employees have job responsibilities which require Web and other Internet applications. Because of the availability of intrusive software and the existence of various motivations, employees are subjected to unsolicited pop-up windows, browser hijacking, unintended release of confidential information, and unwanted e-mail. These intrusions are a significant problem for employees and employers because they waste resources and create liability situations. Solutions examined include education of employees, standards of practice in the conduct of job-related Internet use, policies regarding Internet use for non-work-related purposes, and deployment of protective technologies. Constant attention to evolving threats and updating of the solutions is also essential to successful use of the Internet in the workplace.

INTRODUCTION

Privacy has been defined as "the right to be left alone." Employees sometimes invoke this definition regarding their rights to use the Internet, but another side to it is the interest shared by employers and employees to be protected against unsolicited Web intrusions. Other chapters of this book address the statistics associated with browsing to non-work sites during work hours, from employer-owned computers, and the sending and receiving of personal e-mails. The enormous problems associated with these phenomena are complicated by the uncontrolled proliferation of unsolicited Web intrusions. These intrusions take the form of unsolicited and unwanted advertisements in pop-up windows; hijacking of the browser during the process of legitimate surfing; collection of personal, personally identifiable, and proprietary information without informed consent of the owner of the information; and unsolicited and unwanted email, sometimes with viruses.

The technologies that are used to accomplish these intrusions are known generically as "push technologies," based on their being automatically served up or "pushed" to client computers. By comparison, "pull technologies" make information available when the user makes explicit requests for the information. In the context of any given workplace and any given worker with a job to do, if the Internet is one of the tools available to do the job, it must be expected, in today's Internet environment, that the employee will encounter unsolicited Web intrusions.

The purpose of this chapter is to arm employers and employees with the necessary analytical tools to establish appropriate protections so that these push technology intrusions: (1) do not create time, bandwidth, and other resource wastes which are unacceptable to employees and employers; (2) do not create the potential for unfounded charges of inappropriate use of work time or other resources; (3) do not hamper the employee's ability to do the job; and (4) do not permit activities which would subject the company or the employee to liabilities for activities beyond their control. While the technologies are likely to change, policies and practices can be developed and implemented so that risk exposure on the part of both employers and employees is quite limited.

THE TYPES OF INTRUSIONS

Four types of intrusions are prevalent in the Internet world of today. First is the intrusion of unsolicited, non-relevant pop-up window advertisements (Frackman, Martin, & Ray, 2002). These windows are generally sent to a local workstation when the user links to a site that has contracted to provide the vehicle (usually a legitimate IP address) for pushing the advertising to a potential customer. Some of these are the result of some analysis and targeting based on data collected by or through the linking site, but many are simply pushed to all users.

A second type of intrusion is the spurious collection of personal, personally identifiable, and proprietary information. This type of information collection could include surreptitious collection of any data stored on a computer that is connected to the Internet (Frackman, Martin, & Ray, 2002; Spitzer, 2002). In addition, data unrelated to a given interaction or transaction are often requested, and sometimes even required, to be entered by the user in order to access the needed website. Among the many uses for information collected in this way is the generation of intrusive advertising windows and advertising spam e-mails. Data collected in these ways are often combined into databases and sold or used repeatedly in ways the unsuspecting user has no knowledge of.

Intrusions are also created when products called "scumware" change the appearance of Web pages that are being browsed (Bass, 2002). The link to this type of software is often under the guise of a free service or utility that is going to make something the user wants to do easier or better (Tsuruoka, 2002). But the reality is that scumware floats pop-up ads over other content, inserts its own hyperlinks into a user's view of a Web page, and reroutes existing links to unauthorized sites (Bednarz, 2002). Many times these changes are simply inconvenient to the user in terms of dealing with multiple windows, but other difficulties arise frequently, including attempts to communicate outside the firewall and difficulties in accomplishing simple close-window operations.

The final type of intrusion relates to unsolicited e-mail. Unsolicited e-mail is often generated when the e-mail address is used in some public forum such as a chat, instant message, or a game site or when it is harvested by scumware, spyware, sniffers, snoopers, and similar software products (Credeur, 2002). E-mail addresses are also shared and sold by many Internet page owners who might have collected the information for a purpose and find there is a market for their database of addresses. Unsolicited commercial e-mail is commonly known as "spam." Other sources of unsolicited e-mail include mailing lists of friends, relatives, coworkers, and outside business associates who broadcast messages of humor, inspiration, human interest, or personal activities or perspectives (Retsky, 2002). Finally, e-mails are generated by software that either results from the activity of a virus or carries a virus capable of infecting the recipient's computer.

THE PROBLEM WITH INTRUSIONS

Knowledge workers and other employees who make up today's workforce are expected by their employers to accomplish more and more in the work time they have (Simmers, 2002). Employer expectations are rising and competition is keen. Quality employees strive to maintain job focus, to stay on task, and to perform their jobs efficiently. Intrusions which create workplace situations where employees are distracted, threatened, or slowed down in the performance of their job responsibilities are not welcome by either employer or employee.

Workplace intrusion issues are addressed by a wide variety of efforts to provide a safe, secure, pleasant work environment. Policies and regulations are widely utilized to guard against workplace violence and harassment, and to minimize physical distractions and annoyances. Many workplaces have standards related to telephone usage, smoking, noise, visitors, and peddlers. Workplaces establish security through a variety of measures beyond policies and standards. These security measures rely on restricted entry to certain buildings, floors, and rooms, through the use of various forms of identification screening, locks, schedules, registration, and guards.

In organizations with some dependence on the Internet for performance of employees' job duties, whether these involve electronic commerce, electronic business, research, individual productivity, or enterprise wide systems, the need for protection from intrusions, threats, and distractions in the Internet world parallels the physical world (see Table 1). Responsible employers and employees have a duty to make those protections as routine in the Internet world as they are in the physical world for several reasons. First, employees

Table 1. Intrusion Parallels in the Physical and Internet Worlds

Types of Intrusions	
Physical World Intrusions:	**Internet World Intrusions:**
Unauthorized Personal Visitors	Personal E-mail Pop-up Windows
Vendors	Pop-up Advertisements Spam E-mail
Competitors	Spyware Snoopers
Vandals	Hackers Viruses Trojan Horses
Thieves	Hackers Scumware Spyware Sniffers
Advertisers	Pop-up Advertisements Spam E-mail

need to not be diverted from their job duties reading unsolicited e-mail; identifying, quarantining, and removing viruses; closing unsolicited pop-up windows; escaping from hijacked-browser links; conducting searches to assure that their personal information is not being shared; and sending opt-out notifications related to proprietary information (Simmers, 2002; Retsky, 2002). These activities should be viewed as wasting resources by taking employee time, adding traffic to the network, using up bandwidth on the network, and clogging hard drive and other secondary storage space on company computer systems (Credeur, 2002; Privacy Agenda, 2002; Hillman, 2002).

A second reason that intrusion protections should be routinely utilized in the workplace relates to protection from hostile work environments. Harassing and otherwise undesirable speech, displays, and behaviors are unacceptable in the physical workplace, but in the Internet workplace it is easily possible that undesirable images and written communication can appear on computer screens, in e-mails, and on hard disks and other secondary storage media through no fault of the computer user (Simmers 2002). These might take the form of hate messages, pornography, highly personal products and services, games, and casino advertisements (Bass, 2002). An employee who receives such messages might individually feel threatened, annoyed, embarrassed, harassed, or insulted.

Further, if a co-worker, employer, or customer were to encounter such messages or images on the employee's computer display or in the employee's computer file storage, it could be erroneously assumed that the employee participated in or was interested in the content. Such communications are often regulated in acceptable use policies of companies and in personnel handbooks. Employees could be subject to harassment or inappropriate conduct charges, or an employer could be held liable for such conduct even though the communication had been initiated outside the employee's control (Simmmers, 2002).

A final major reason for establishing protection from Internet intrusions involves the protection of individual personal and corporate proprietary/confidential information. When the Internet is used for many types of work-related activities, data contained in corporate databases, log files, and password information are vulnerable to unauthorized, surreptitious retrieval. Employees are thereby exposed to accusations of divulging confidential information, and companies risk loss of competitive advantage and loss of customer goodwill. This type of intrusion is more prevalent in situations where the computer has a static IP address or is "always on" or connected to the Internet. Outsiders use software that will identify the live IP address and make connection, then proceed to retrieve unprotected information without the knowledge of the user or owner. Once the retrieval process is completed, no record of the transfer exists on the owner's machine and no control exists concerning the disposition of the retrieved information.

SOURCES OF INTRUSIONS

Advertisers, hackers, scammers, private investigators, and government agencies all have motivations to learn as much as they can about Internet users in general and about specific Internet user activities and habits. Advertisers and their agencies must get their product or service information to potential customers (Tsuruoka, 2002). Hackers and scammers are interested in pushing their abilities to gain access, sometimes to wreak havoc, other times to take advantage (*Consumer Reports,* 2002). Private investigators and government agencies have new surveillance challenges because of the Internet.

For each of these situations, two events need to occur: the intruder must learn how to identify the "target" computer, and the intruder must establish a communication with the "target" computer. The communication might be in the form of sending an e-mail or pop-up window directly, or it might involve monitoring keystroke or mouse click activities, reading stored data, or modifying messages sent to the target browser by other computers.

For the purpose of identifying the target computer, a variety of techniques and technologies might be utilized (Privacy.net, 2002). The two primary types of addresses are e-mail addresses and IP addresses (with or without the associated domain names). These addresses are available directly through a wide variety of listings and services, some of which users have willingly subscribed to, some of which users inadvertently or unwittingly participate in, and some of which are collected in clearly surreptitious ways that users must go to great pains and sometimes expense to avoid (Credeur, 2002). In addition to listings that are available or created by third parties, intruders sometimes generate addresses and send probing messages, looking for an active target computer and a response (Raz, 2002). These addresses might be constructed randomly or use patterns composed of frequently used names, words, or other standard addressing combinations (Frackman, Martin, & Ray, 2002). Both IP addresses and e-mail addresses are used in this type of probe.

Internet users are often unaware of the intrusive capabilities of Internet technologies and the behaviors that permit the intrusions to occur. In addition to Web surfing through a browser, many Internet users routinely participate in chat sessions; play online games; register for prizes; respond to offers for free software and services; and register preferences for news, sports scores, stock quotes, music, entertainment, credit checks, and other seemingly innocuous elements. Furthermore, Internet users often search the Web for medical advice, financial advice, career advice, and the like — never suspecting that someone along the way might begin tracking the clicks for the purpose of targeting advertisements, profiling the user, or conducting surveillance activities. Any of these activities subject the target computer to intrusions such as pop-up window advertisements, click tracking, data retrieval, and browser hijacking (Bednarz, 2002).

Software and service providers are readily available to accommodate the needs of individuals and companies who wish to collect information from and about Internet users including their personal habits and data (Spitzer, 2002). Many of these software and service providers are using the same technologies that companies use to track the online activities of their employees. And even in work-related use situations, Internet users are often trapped into giving personal information in exchange for the ability to access needed sites. Once given, this information — without context, consent, or verification — is often sold, used for other purposes, mined with other data to create profiles, or used directly for targeting advertising pop-up windows or e-mails (Credeur, 2002). The result can be that unexpected, unsolicited, and unwanted messages can appear on an employee's computer screen or in an employee's e-mail, or the employee's browsing can be interrupted because scumware has hijacked the browser and provided links to sites other than those that were intended and appropriate.

WEB INTRUSION PROTECTION STRATEGIES

Protection from intrusions in Web-related activities is important for both employee and employer. Moreover, successful protections require that employees and employers become active partners in the ongoing venture. Protection against intrusions is not accomplished by applying a static, one-time fix and expecting that no further attention is required. A routine process for reviewing intrusion threats, and updating technologies and practices is essential if a workplace is to be successfully protected against undesirable intrusions.

From the standpoint of the employee, each person should exercise care and maintain a watchful eye in all Internet communication processes (Tynan, 2002). Employees are responsible for understanding and observing the Acceptable Use Policies of their employers. Further, employees should be aware of where vulnerabilities are likely and should act in ways that are protective of the company's data and network resources. How these behaviors are implemented and the details of specific implementations need to be governed by the type of job the employee is doing, and the corporate culture and policies regarding employee use of the Internet.

Employees should be given guidance in both the policies regarding Web use and the safeguards that the company has put in place. Employees should also be given information regarding the types of intrusions to watch for and the corrective or protective measures that can be implemented in the event of an intrusion (Tynan, 2002). Employees should also be warned about the types of activities that invite, or at least facilitate, some types of intrusions. Depending on the work environment, job responsibilities, and skill level of employees, employers might incorporate information concerning protections against Web intrusions in routine training sessions or staff meetings, newsletters, occasional e-mail reminders, or FAQs on a website. Employees should utilize all available software options and settings as efficiently as possible to prevent unwanted intrusions while maintaining the ability to do the job efficiently. This balance is often difficult to achieve and might require technical support for effective implementation in individual cases.

Employers seeking protections from unsolicited and unwanted Web intrusions are obligated to establish a safe work environment by installing protective measures on the company's networks. Anti-virus software is an essential component of any Internet e-mail system, and can easily be purchased, installed, configured, and updated regularly. While not absolute in the protections that these packages provide, they are of high enough quality that no computer should be given Internet e-mail access without a good, active, updated anti-virus program. Computers and networks that contain sensitive, confidential, or proprietary data; customer data; credit card numbers; access codes; passwords; or employee personal data must be protected by one or more firewalls. Other possibilities for protections include anti-spam software, e-mail filters, and high security operating system privacy settings (Frackman, Martin, & Ray, 2002). Careful analysis of the specific job requirements is often necessary to properly implement many of these protections. Additional com-

Table 2. Physical and Technological Protections in the Physical and Internet Worlds

Physical World		**Internet World**	
Intrusions:	**Physical Protections:**	**Technological Protections:**	**Intrusions:**
Unauthorized Personal Visitors	Fences	Acceptable Use Policies; Passwords	Personal Unsolicited E-mail; Pop-up Windows
Vendors	Locks	Pop-up Blockers; Filtering Software	Pop-up Advertisements; Spam E-mail
Competitors	Guards	Firewalls	Spyware; Snoopers
Vandals	Identification Systems	Anti-virus Software	Hackers; Viruses; Trojan Horses
Thieves	Surveillance Systems	Firewalls	Hackers; Spyware; Sniffers
Advertisers	Admittance Policies	Filtering Software	Pop-up Advertisements; Spam E-mail

plications arise if the corporate network allows remote access by employees and older technologies like FTP and Telnet. Finally, many companies should establish standards of practice regarding responding to unsolicited e-mails, registering for miscellaneous online services, opting-out of service offers and spam messages, forwarding of chain e-mails, and providing personal information that seems unrelated to a given transaction or job duty, because many of these actions will result in more, not less intrusive traffic (Clark, 2002).

EXAMPLES OF CURRENTLY AVAILABLE PROTECTION TECHNOLOGIES

Just as there are physical protections from intrusions into offices and factories, technological protections protect from intrusions in the Internet world (see Table 2). Various technologies are available to assist in the protection against unsolicited and unwanted Web intrusions. EPIC's Online Guide to Practical Privacy Tools (Electronic Privacy Information Center, 2002) contains a comprehensive and reliable set of technology tools and reference links to test vulnerability and protect network computers. Recommended technologies include anti-virus software, e-mail client settings, hardware and software firewalls, anti-spam software, operating system privacy settings, and anti-scumware software (Bass, 2002; *Consumer Reports,* 2002). Options exist for deploying these technologies at the individual workstation level, local area network server level, or Internet gateway level. In networked environments, these might need to be deployed at multiple locations between the individual workstation and "the Internet."

In practically all cases, anti-virus software should be running on every e-mail client, and detailed attention should be given to all of the filtering and privacy options on the e-mail client. Privacy settings available on the local operating system should always be set as high as possible, given the constraint of needing to get the individual's job done.

In many cases a local area network can operate behind a firewall that will provide protections from snoops, probes, sniffers, and spyware. Often a separate firewall is needed on each individual workstation in addition to the one associated with the LAN server. And in the case of multiple LANs sharing access to the Internet through a single gateway, it might be necessary that another firewall be installed at the gateway level.

Examples of anti-virus software include Norton and McAfee anti-virus software. These programs contain databases of virus definitions that must be updated regularly. The programs scan all system areas for viruses, worms, and other identified program code that could modify contents of the system or cause undesirable activities like spam e-mail, or otherwise wreak havoc with the computer system or tie up system resources. If problematic code is identified, the code is quarantined or repaired and the user receives a report.

Personal firewalls are typically software firewalls. Personal firewalls include Norton Personal Firewall, McAfee Firewall, and ZoneAlarm Firewall. Corporate firewalls usually combine hardware and software. CheckPoint Firewall, Raptor Firewall, and Gauntlet Firewall are examples of corporate firewalls. Through the use of firewalls, hackers are prevented from breaking into the system. Further, when a software firewall is running and properly configured, programs on the computer cannot connect to the Internet without the user knowing about it, and data cannot be sent out without the user knowing about it. Firewalls operate based on a set of rules established by the user (Bednarz, 2002).

Examples of anti-spam software include MailMarshal, Spaminator, SpamMotel, and SpamEater (Clark, 2002). This type of software can compare received e-mails with the user's e-mail address book and can also review an existing extensive list of known spammers (these spams might be deleted by the software). Another capability of anti-spam software might be to scan the subject heading and the content of the e-mail to detect spam (Clark, 2002). If desired, anti-spam software usually can provide a junk mail folder from where the user can scan the e-mails personally.

Examples of Windows 98/2000 operating system privacy settings include Internet option security features where the users can set the security level by setting different options such as whether to accept/deny ActiveX controls, cookies, etc. Also, the user can add digital certificates and website ratings for safe surfing. Windows XP: Home Edition has built-in Internet Connection Firewall software. Windows XP Professional Edition has security management features in addition, such as encryption.

Examples of anti-scumware include Lavasoft's free Ad-aware, Symantec's new Client Security (intrusion detection software for corporations), and Zone Labs Integrity line of software products (Bednarz, 2002). These programs scan the local computer components for known spyware and scumware in much the same way that virus software scans files before they are opened. Any offending programs are removed, or otherwise made non-functional.

Table 3. Behavioral Protections Against Web Intrusions

Employee Practices to be Encouraged Through Training and Policies	
DO:	**DO NOT:**
Update virus software frequently and regularly	Play online games
Establish high security browser settings	Unnecessarily engage in open chat sessions
Read privacy statements critically	Participate in online auctions
Minimize use of general browser searches	Reply to unknown e-mails offering to remove you from lists
Set filtering software appropriately for the environment	Send chain e-mails that make promises of rewards or threats of doom
Utilize as many features of firewalls as possible	Sign up for sweepstakes and give-aways in exchange for unsubstantiated future benefits
Clear cookie files, log files and other temporary files frequently	Provide personal information to unknown parties
Update anti-scumware software and pop-up window protections frequently and regularly	Provide personal information that is not relevant to a transaction or relationship to known parties

EXAMPLES OF INTRUSION PROTECTION PRACTICES FOR EMPLOYEES

In addition to technological protections, behavioral strategies can be incorporated into an organization's unsolicited Web intrusion protection strategy (see Table 3). Employees should be instructed through whatever communication format the company uses to adhere to certain practices regarding protection of the company's network resources. These instructions might be part of an employee handbook, part of the Acceptable Use Policies associated with the Internet, discussed at staff meetings, included in electronic or paper newsletters, or presented at orientation sessions and workshops. Instructions should provide ways to assure that the company is not put at risk through loss of proprietary or confidential information; through display, broadcast, or storage of objectionable materials; or through loss of employee time and other company resources because of browser hijackings, virus attacks, pop-up windows, or unsolicited e-mail (Simmers, 2002; Siau, Nah, & Teng, 2002).

Individual Web behaviors which are likely to result in unsolicited communications include open chat sessions, online games, auctioning, and dashboard news services (Crouch, 2002). Corporate Acceptable Use Policies should address the appropriateness of these activities in the workplace (Siau, Nah, &

Teng, 2002). Individual jobs should be assessed to determine if these activities are essential or desirable for an employee to fulfill their job duties. Expectations regarding this type of activity should be clearly communicated to each affected employee. Siau, Nah, and Teng (2002) provide a useful set of guidelines for writing acceptable Internet use policies.

Employees should be instructed concerning the protection of any information the company considers proprietary or confidential. Specific procedures should be established to protect this information. Again, expectations concerning how information is to be protected and what information is to be protected need to be clearly communicated to employees (Frackman, Martin, & Ray, 2002).

Employees should also be instructed in the ways that are used to collect live IP addresses or live e-mail addresses under the guise of providing a service or providing an opt-out option for an unwanted newsletter or other "service" (Frackman, Martin, & Ray, 2002). Employees should also be advised against participating in online drawings, lotteries, and other games of chance promising the potential to win valuable prizes. Just the act or responding can activate intrusive communications, and many times the participant is asked for personal information that can be used for further intrusion.

Similarly, users are often tempted to reply to spam e-mails that provide for unsubscribing or opting out of further communications. These are frequently used as a guise for validating the e-mail address so that the user will then receive more, not less spam e-mail (Clark, 2002; Porcelli, 2002). Users in reasonably well-protected environments will tend not to get a large number of this type of message, but should have periodic reminders of the hazard.

Care in opening e-mail attachments of unknown origin is a widely understood guideline. Viruşes and Trojan horses are promulgated through e-mail attachments. Some of the more notorious ones manage to be masqueraded so that they are undetectable for a time by virus-detection software. All organizations should have a procedure to remind employees of this hazard and of the need to resist the temptation to open files attached to e-mails of unknown origin, no matter how enticing or sincere the message or subject line might sound.

If a job requires heavy use of a wide variety of commercial websites and acceptance of cookies, the employee should be aware of the repercussions of such activity and should periodically review and delete temporary files and folders, unneeded cookies, and history files (Bass, 2002). Further, employees using this type of browsing need to pay close attention to opt-in and opt-out choices, and exercise care in the use of those options (Tynan, 2002; Frackman,

Martin, & Ray, 2002). When e-mail addresses are frequently required to be provided to access online sites and services, it is useful to maintain a separate e-mail address for that purpose and use another official e-mail address for correspondence related to internal company matters.

SUMMARY AND CONCLUSIONS

The problem of unsolicited and unwanted Web intrusions is multi-faceted. It includes unwanted communications that take employee time, network resources including bandwidth, and storage. Some communications through e-mail might be offensive to individuals, damaging to computer systems, or damaging to the company's ability to provide services. Communications through the Web and other channels likewise can be offensive or create service slowdowns. They can also collect information that is used for undesirable or unauthorized purposes. The net result is lost company revenues, increased costs, and potential for liability.

The solutions to the problem of unwanted and unsolicited Web intrusions involve a multi-faceted array of technological protections, employer policies and standards, employee practices and training, and routine review of the solution set to identify needed improvements. The technologies need to be deployed at a variety of levels within the network structure and take into account the specific job needs and the corporate culture. The solutions need to be applied in the context of a partnership between the employer and employees so that when new intrusions are identified, resolution can be achieved with a minimum of disruption in the work flow. Further the deployment of technological solutions needs to take into account the impact that it has on an employee's ability to successfully complete the assigned job duties, with a minimum of encumbrances.

The use of the Web and other Internet-enabled technologies are important tools for many companies and employees. The abuse of the system by those outside the system must be addressed in a positive, collaborative effort to minimize as many risks as possible. Successful companies and their employees will find best advantage of the Web technologies when they work together to solve the problem of unwanted and unsolicited Web intrusions.

REFERENCES

Bass, S. (2002). Steve Bass's home office: Beware: Sleazy websites, spyware: Underhanded websites, spyware, and how to protect yourself from them. *PCWorld.com,* (March 13).

Bednarz, A. (2002). Critics decry spread of 'scumware' on the Web. *NetworkWorld,* 10(33), 1 & 62.

Clark, B.L. (2002). You've got too much mail. *Money*, (June).

Consumer Reports. (2002). Cyberspace invaders (June).

Cradeur, M.J. (2002). EarthLink wins $25 million lawsuit against junk e-mailer. *Atlanta Business Chronicle,* 25(16).

Crouch, C. (2000). The Web inside outlook 2000. *PCWorld.Com,* 11(April).

Donlan, T.G. (2002). Editorial commentary: Slicing spam. *Barron's,* 82(27).

Electronic Privacy Information Center. (2002). *EPIC Online Guide to Practical Privacy Tools*. Accessed September 29, 2002, from: http://www.epic.org.

Engst, A.C. (2002). Stop spam! *MacWorld,* 19(8).

Frackman, A., Martin, R.C., & Ray C. (2002). *Internet and Online Privacy: A Legal and Business Guide*. ALM Publishing.

Porcelli, N. (2002). FTC settles first spam cases. *Intellectual Property & Technology Law Journal,* 14(6).

Privacy.net, the Consumer Information Organization. (2002). *Being Traced Over the Internet.* Accessed September 29, 2002, from: http://www.privacy.net.

Raz, U. (2002). *How Do Spammers Harvest E-Mail Addresses?* Available online at: http://www.private.org.il/harvest.html. [Referenced in Engst, A.C. (2002). Stop spam! *MacWorld,* 19(8).]

Retsky, M.L. (2002). At least one firm's willing to sue spammers. *Marketing News,* (April 29).

Siau, K., Nah, F.F., & Teng, L. (2002). Acceptable Internet use policy: Surveying use policies of three organizations — educational institutions, ISPs and non-ISPs. *Communications of the ACM,* (January), 75.

Simmers, C.A. (2002). Aligning Internet usage with business priorities: Regulating Internet activities so that targeted outcomes remain within acceptable limits. *Communications of the ACM,* (January), 71.

Spitzer, E. (2002). *Major Online Advertiser Agrees to Privacy Standards for Online Tracking.* Press Release, Office of New York State Attorney General, August 26.

Tillman, B. (2002). Spamming gets a closer look. *Information Management Journal,* (March/April), 10-15.

Tsuruoka, D. (2002). Yahoo marketing pitches becoming very personal means of boosting revenue. *Investor's Business Daily,* (July 25).

Tynan, D. (2002). How to take back your privacy. *PC World,* (June). Accessed September 29, 2002, from http://www.PCWorld.com.

Chapter VIII

Monitoring Strategies for Internet Technologies

Andrew Urbaczewski
University of Michigan - Dearborn, USA

ABSTRACT

Managers are faced with many decisions regarding monitoring. For an electronic monitoring effort to be successful, it is important to match the correct monitoring strategy with a complimentary monitoring technology and implementation. This chapter lists many of the potential goals for monitoring, strategies to accomplish those goals, technologies which match the strategies, and implementation plans. Managers can consult this chapter to assist in ensuring that unintended effects do not occur from a haphazard approach to electronic monitoring.

INTRODUCTION

Most large organizations that are providing Internet access to their employees are also providing some means to monitor and/or control that usage (Reuters, 2002). Many other chapters in this text are devoted to different aspects of human resource management in monitoring personal Web usage. This chapter is designed to provide a classification and description of various control mechanisms for the manager who wants to curb or control personal Internet usage in the organization. Some of these solutions will be technical, while others are social solutions, relying on interpersonal skills rather than the hammer of the logfile to curb cyberslacking.

First, this chapter will discuss the goals for the monitoring program. Second, a list of different activities to monitor and/or control will be provided. Third, a discussion of different techniques for monitoring will be explored. Fourth, a review of several technical products will be provided. Finally, the chapter will end with a discussion of fit between corporate culture and monitoring.

GOALS FOR MONITORING

Why do companies monitor their employees? Organizations do this for a variety of reasons, including simply "because they can." An electronic monitoring effort is often difficult to establish and to maintain, so before an organization would begin such an effort, there should be clear goals for the monitoring.

The popular press is filled with articles of employees frittering away time on the Internet (Swanson, 2002). In the beginning, employees were likely to spend unauthorized time on the Internet at pornography and gambling sites, but now news and online shopping are likely to be found on the screen of the cybersloucher (Reuters, 2002). This is in stark contrast to what employers had sought when they implemented Internet connections.

In response to these challenges, employers often created acceptable use policies (AUPs) which told employees for what they could and could not use the company Internet connection. Some organizations already had AUPs implemented to keep games and other frivilous computing technology outside of the organization, and they simply modified these policies. For many other organizations, including new organizations, they had to create new AUPs which addressed the Internet threat to productivity head-on.

As any parent would know, policies are useless without any enforcement. However, in today's litigious society, it behooves the accuser to be absolutely certain of a transgression before acting to enforce the policy. Monitoring tools sought to identify and record unauthorized Web usage in an irrefutable log which can stand as evidence at a hearing. It is often hoped that the mere threat of punishment will get employees back on task, often with some success (Urbaczewski & Jessup, 2002). Below are listed a few potential goals of a monitoring effort.

Increase Employee Productivity

The Internet was introduced into many organizations as a tool to aid employees in completing their jobs more efficiently. This was the case with the introduction of other information technology tools like spreadsheets and accounting packages. These tools provided few opportunities for the employee seeking to slouch on employer time. Employers were much more concerned with employees loading games onto their desktop PCs which may contain potentially harmful viruses or otherwise disturb the computing atmosphere within the organization. Plugging into the Internet was an entirely different issue for employers, as the PC was now in many cases an electronic equivalent of a water cooler, break room, or smokers' perch at an organization.

The Internet can indeed be a place where employees spend enough time that it cuts into their productivity. To curb this potential problem, an organization could implement a monitoring program which records the amount of time spent at non-work-related Internet sites, or potentially even block access to all of these sites. An alternative may even be to limit access to frivilous sites to non-production hours, such as before or after normal working hours or during a standard lunch break.

Bandwidth Preservation

In some organizations, monitoring is established for reasons other than the direct productivity of workers. Rather, the issue is that all of the network bandwidth, or capacity for carrying information, is being used by applications and instances that are not directly related to the organization's goals. This is often due to people listening to streaming audio or watching streaming video, which is a constant drain on the bandwidth. People can also engage in excessive uploading and downloading of files across the networks, but this can also be the result of poorly coded programs that are unnecessarily "chatty." Constant network activity reduces its performance considerably.

If bandwidth usage is the problem, there are two general solutions: purchase more bandwidth, or limit the usage of existing bandwidth. Monitoring programs can be established when an organization seeks to limit the usage of existing bandwidth. A variety of technical solutions, as discussed later, can help the organization to preserve existing bandwidth.

Legal Liability Reduction

Along with productivity and bandwidth usage, organizations are also concerned about Internet usage from the potential exposure it brings to legal liability. Consider the following fictitious scenarios:

> *"Organization X today was sued for negligence, as an employee was running a child pornography server inside the corporate network."*
> *"ABC Corporation today was sued by a former employee who is now in treatment with Gambler's Anonymous. He is charging that ABC, by placing an Internet terminal on his desktop, essentially gave him unfettered access to the virtual casinos thriving on the Internet."*
> *"Company B is defending itself today against a privacy lawsuit. It is charged that when an employee downloaded a file-sharing program, that program was equipped with a backdoor which allowed malicious hackers entrance into Company B's networks. These hackers then took thousands of credit card numbers and personal data from the databases..."*

The above scenarios are scary enough, but consider additional possibilities like sexual harassment suits and industrial espionage, and the legal risks mount. Organizations indeed may wish to monitor Internet connections to prevent any potential legal liabilities from allowing illegal activities to be conducted on their networks. A monitored network that then does not catch certain activities may then also leave the company with legal exposure for not conducting the monitoring correctly, so consultation with the corporate legal department should be conducted before engaging in monitoring for the prevention of legal risks.

ACTIVITIES TO MONITOR

While the title of this text alludes to the process of personal Web usage, there are many non-work-related activities that an employee can do with an Internet connection that are not necessarily Web related. This section will first look at Web activities, but also many other methods of cyberslacking.

Web Surfing

While the Web gave a pictoral interface to the Internet that made business-to-consumer (B2C) e-commerce a reality, it also provided a gateway for workers to fritter away the normal work hours. The Web provides a smorgasbord of information that helps knowledge workers on a daily basis, but it also provides access to shopping, gambling, pornography, hate speech, games, sports, and news, just to list a sample of possible topics. As the rest of this text deals with Web usage specifically, the other details on this area will be left to other chapters. Monitoring Web surfing would be an obvious choice for any of the aforementioned goals for establishing a monitoring program, but productivity concerns seem to be the biggest reasons an organization would monitor Web usage.

E-Mail

E-mail was the first "killer app" for the Internet, and thus it has been around much longer than the other technologies discussed in this chapter. E-mail communications have become very commonplace in business today, and it is almost as common to see an e-mail address on a business card as it is to see a telephone number on one. Some individuals have been known to e-mail friends and family for a majority of the workday, right from their desks. This is perhaps a 21st century version of the 20th century worker who talked on the telephone continuously to friends and family. E-mail can thus be monitored to ensure that individuals are not using the technology for personal use instead of furthering business communications.

However, personal productivity is not the major reason that most organizations would implement e-mail monitoring. Rather, e-mail monitoring is used for bandwidth control and reduction of legal liabilities. The simple e-mail message that one might send to a colleague discussing a topic does not eat up much bandwidth. What does use a lot of bandwidth are large attachments to e-mail messages. Sometimes these attachments are perfectly within the boundaries of normal work, like a large work document or a presentation that is being discussed between colleagues. However, these are often large animations or pictures that are intended to be lighthearted. While they may be humorous to the recipients, they often can clog e-mail servers and enterprise networks, as when a worker e-mails a five megabyte animation to 200 of his/her closest friends inside and outside of an organization. Sometimes these humorous applications can also come with destructive virus payloads, or they actually may only be worms and viruses intended to appear like jokes or pictures, such as the Anna Kournikova worm that was popular on the Internet in 2001 (Fonseca, 2001).

Companies may also monitor e-mail to prevent exposure to legal issues. An employee may be using corporate assets to further a personal business. If it appears that this usage has gone unchecked, the organization could potentially be found liable for any of the wrongdoings of that employee. A court could find that the organization became an unwitting, but yet still willful, silent partner in the employee's activities. The waters get even murkier when the business is on its face illegal, such as running an illegal bookmaking ring, trading in child pornography, or even facilitating software piracy. Moreover, e-mail can be the tool of the disgruntled employee, who uses the system to send out trade secrets or other important internal information to individuals outside the organization. E-mail monitoring can help prevent all of these transgressions.

File Sharing

Many network managers today rue the day they first heard the word *Napster*. Since then, the Internet has been awash in the peer-to-peer (P2P) networking technologies designed to send files between users. While the P2P blitz has in many ways been a boon for companies, and Notes creator Ray Ozzie founded Groove Networks as a corporate messaging aid using P2P technologies, it has also been the main tool used by those wishing to trade copyrighted digital materials.

No matter which of the three aforementioned goals of monitoring an organization might be seeking, file-sharing would be one of the most important applications to monitor. First of all, employees may spend large amounts of time each day looking for certain songs, movies, or software applications available from other P2P servers. Second, these files are often very large, from several megabytes to several hundred megabytes. Especially when it is done by more than one person, file-sharing can cause a serious strain on bandwidth. Furthermore, a number of viruses and worms have been found to exploit the P2P technologies as one more way to gain entry into a corporate network. Finally, the legality of copyrighted file-sharing is murky at best and often found to be explicitly illegal. If a corporation is found to permit or turn a blind eye to such activities, it could be found legally liable for aiding and abetting the activities.

Instant Messaging

Instant messaging (IM) has become another popular means for communications. While the phenomemon first became popular in the mainstream with ICQ in the mid-1990s, similar technologies have existed since the 1960s with timesharing systems. Now some organizations are even starting to offer IM as a means for customers to communicate with technical support people and gain other types of real-time assistance.

IM is generally not a drain on bandwidth, although there are some experimental worms and viruses that spread through IM mechanisms. When IM is noted as a problem in an organization, it is usually for the same reasons as e-mail. Some individuals arrive at work in the morning, open up their IM programs, and then chat with friends the rest of the day, much as some workers did (and still do) with the telephone. IM also has some legal liability concerns, depending on the organization. Many of these are consistent with those for e-mail, but one feature that has plagued IM is an inability of the "conversations" to be recorded and logged. This "feature" is important in industries where all conversations must be recorded by law, such as the securities industry. Technologies are just starting to appear which will capture the content of IM programs, but this is monitoring in and of itself.

DIFFERENT MONITORING STRATEGIES

There are several different control mechanisms that an organization might use, but they are generally grouped into one of two categories: managerial and technical. The managerial techniques for monitoring are similar to ways that monitoring of employees has been done for decades: walking around and keeping one's eyes open. Managers may not feel a need to monitor an employee that is not causing any problems, i.e., work is getting done on time, no complaints about collegiality, etc. When a manager starts to wonder about an employee's performance or collegiality, that might be when the manager starts to pay more attention to that employee's work habits. Dropping by his/her office to see what the employee is doing on a routine basis is a start. Is the computer always the focus of attention? Does the employee hurriedly close windows on the screen when the manager shows up? Is a door that used to always be open now frequently closed? Are there abundant printouts of computer pictures surrounding the employee's desk? Is the employee always burning CDs or other media? These are all definite clues to how an employee may be using an Internet connection.

By and large, however, the more popular means of monitoring employees is through technical solutions. In many ways, this makes sense — a technical solution to a problem assisted by technology. Electronic monitoring functions like a "big brother," keeping a watchful eye on all systems in the network at all hours of the day and night (or whatever subset of those systems/hours that a manager may choose to watch). Records can then be kept and offered as "proof" of an employee's misgivings as related to using organizational computing equipment and network time. There are two main ways that an organization can accomplish electronic monitoring of personal Internet usage, and they are discussed in the next section.

ELECTRONIC MONITORING TECHNIQUES

Logging at the Gateway

The Internet functions as a "network of networks." When a computer tries to make a connection to another computer, it first checks to see if the destination is on the same local network (subnetwork or subnet) as it is. If the destination is not on the same subnet, then the packet must be routed outside the network through what is commonly referred to as a "gateway." The router that functions as the gateway is essentially the virtual in/out door from the organization's network to the rest of the world. Many logging technologies are then designed to capture and record all of the packets that enter and leave the organization, or at least the header information that indicates the sender, recipient, and content of the message.

Gateway logging can be a useful tool in that it provides a central point of control for the network. However, it is difficult to accurately gauge how long an employee stares at a particular page, and if all that time (s)he is actually staring at that page or if (s)he has actually gone to lunch and returned later. Moreover, gateway logging can quite often be defeated by the use of encryption tools. A recent case involving the Scarfo family in the Philadelphia organized crime scene was using PGP (a freely available encryption program) to code computer files which contained family business (see McCullagh, 2001, for a more detailed description of the entire case and legal concerns). Gateway logging (in this case, the gateway was the Internet service provider) did the FBI little good in identifying the contents of the messages, even though they had a search warrant. Another technology had to be used to get the information they desired, as is discussed below.

Sniffing at the Client

When gateway logging is not sufficient, another means of electronically monitoring connections is to monitor them at the source, or make a record at the client's machine. In the Scarfo case, the FBI did exactly that. They installed a keystroke logging program (whether it is hardware or software, and exactly how it got there, is still classified) on Scarfo's computer. It recorded all of the keystrokes that he used, including the ones that made up his passphrase (a series of words used in PGP, much longer than a password). Once the FBI had his passphrase, they could decode his messages and then had the evidence to make the arrest.

Client sniffing programs are excellent at recording exactly what the user is doing with the computer at any given time. Many will not only record all of the keystrokes that the user makes, but also will calculate mouse movements and active windows, allowing the reconstruction of the entire computing session. Moreover, they capture undesirable activity that may not be directly network related, such as playing games and typing job application letters. However, these programs are not without their own faults. First of all, the manager must install the program on the user's computer, which may not be as easy as it sounds, especially with laptop and other mobile computers. Second, the program must not be detectable (and thus able to be compromised) by the monitored employees. Third, the program must work on a variety of operating systems, including different flavors of Windows, Unix, Linux, and Macintosh, in order to work with all computers. This is not a limitation of gateway logging, as network protocols such as TCP/IP tend to be device independent. Next, the manager has to actually get access to the data captured by the program. Finally, the manager must be able to sift through the mountains of generated data to determine whether or not there is any untoward activity, or enough of it to

Figure 1. Monitoring Strategies Present a Tradeoff Between Information and Intrusiveness

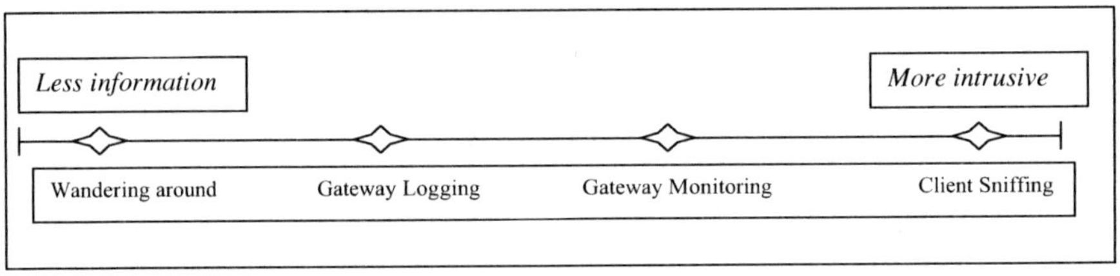

warrant further investigation. This all being said, there are products available which meet the above concerns to varying degrees, and the next section will discuss some of those products.

SOFTWARE PRODUCTS FOR CONTROLLING INTERNET USAGE

As mentioned above, there are various products available to serve as control mechanisms. They are grouped below into four categories. Note that software products come and go, and the availability of certain products and companies are subject to market forces. The categories themselves should remain stable for the time being, although who knows what the future may hold. For example, if this chapter was being written in 1999, there would likely be no section on file-sharing. Table 1 provides a listing of all of the products and a website for the corporation selling that product.

Web Monitoring Products

As the popular press articles have largely focused on employee cyberslacking as a problem with personal Web usage, a number of products have been created to help employers manage these problems and related concerns. The software products are all customizeable to some degree, but there are two main classifications of these products: those that *monitor* and record Web usage, and those that actively *block* access to certain websites deemed inappropriate by the organization. The listing, which is not intended to be exhaustive, details several of these products.

Table 1. Monitoring Products Mentioned in this Chapter

Product	**Website**
Cybersitter	www.cybersitter.com
NetNanny	www.netnanny.com
Websense	www.websense.com
MIMESweeper	www.mimesweeper.com
Elron	www.elronsoftware.com
Message Monitor	www.tumbleweed.com
P2P Traffic Monitor	www.audiblemagic.com
Packeteer	www.packeteer.com
Facetime	www.facetime.com
Vericept	www.vericept.com
Communicator Hub	www.communicatorinc.com
Winwhatwhere	www.winwhatwhere.com
Spy Agent	www.computer-monitoring.com
eBlaster	www.computer-monitoring.com

Cybersitter and *NetNanny* are two programs that are geared largely at the home market, but are used in some smaller organizations. These programs are installed on the client computer and they maintain logs of all the Web pages that are visited by the users. *Websense*, on the other hand, is a tool that is designed to monitor the Web usage of an entire corporate network. It runs on a computer near the corporate firewall, and it logs all Web usage as the requests leave the network. All of these programs can be configured to block and/or record access to certain websites. Some of these programs can be tailored to allow different access rules at different times of day. For example, an organization may wish to allow its employees to use the Internet for shopping and other personal entertainment before and after normal business hours and on the lunch hour but not during the work day. This blocking rule can be enforced by configuring the program in such a manner.

E-Mail Monitoring Products

E-mail can easily be monitored by simply copying the contents of a user's inbox for incoming mail or logging the actions of the simple mail transport protocol (SMTP) server for outgoing mail. However, these logs are often difficult to read, especially when dealing with a large network and a large volume of network traffic. A series of products are made available to help the network manager parse the logs, searching for users or keywords. Some of these products include *MIMESweeper, Elron,* and *Tumbleweed/Message Monitor.*

There are a number of plugins, such as PGP or GPG, for popular mail programs that can send encrypted electronic mail. These plugins are helpful when someone wants to send private information over an inherently insecure network like the Internet. However, they can also deem certain communications unreadable, such as the ones that an organization might be monitoring to control. Logging and reading these encrypted messages is a challenge not easily solved, as brute force attacks on cracking the passphrases can take decades, even with the most powerful computers. Organizations may wish to have separate policies on encryption for monitoring.

Monitoring e-mail sent through popular Web-based providers like Yahoo! or Hotmail can be difficult as well, because the message never passes through the SMTP servers for the organization, nor does the organization have direct control over the user's mailboxes. Monitoring these type of mail services is usually done through a general monitoring tool, as listed in another section below.

File-Sharing Monitoring Products

File-sharing has a history of waxing and waning between one of the easiest applications to monitor to one of the toughest. It some ways it appears that it is almost a game as users of file-sharing services try to devise ways to run their services around and through corporate attempts to halt them. The first file-sharing products, like Napster, always connected to the server on a specific transmission control protocol (TCP) port. Managers were then able to simply block access to that TCP port at the firewall and it would eliminate access to that service. Then a plethora of new services came out using different ports, or the ports were configurable by the users. Some users would simply set their file-sharing traffic to port 80, a port normally left open by network managers as it is used largely to connect Hypertext Transfer Protocol (HTTP) requests. Other problems were created by users demanding that they be allowed to use these programs, especially at high-profile universities like Yale and Indiana. In those cases, something had to be done to limit the amount of bandwidth these services could use, because other legitimate traffic was being crowded out by the file-sharing traffic. A number of hardware and software solutions cropped up to aid network managers in their quest to reduce or eliminate file-sharing traffic.

On the software side, it was mentioned above that already existing firewalls can be configured to block traffic on certain TCP (Layer 4) ports. Other programs, like *P2P Traffic Monitor*, are designed to examine the packets at the application layer (Layer 7) to determine the type of packet and whether or not to block it. Hardware solutions like *Packeteer* plug into the network to control the amount of bandwidth available to certain applications. Packeteer has been most popular at colleges and universities, which in general do not want to be accused of censorship or limiting access to resources, but still have to deal with bandwidth concerns among thousands of users.

Instant Messaging Monitoring Products

IM was one of the toughest applications to monitor for a long time. The nature of IM messages could be likened to "fire and forget," as they behave almost as random packets through a network. The problem was exacerbated because the employers generally did not control the IM servers or the clients. It was in 2002 that applications were created which successfully monitor IM applications and content. These applications were implemented largely to comply with certain U.S. Securities and Exchange Commission (SEC) requirements about keeping logs of all transactions between brokerage houses and their customers. As IM is generally not a big bandwidth hog, it is usually not monitored to conserve bandwidth. IM can become a productivity drain if a person is spending a considerable amount of time each day chatting with friends and colleagues.

Facetime is probably the leader in monitoring IM communications within an organization. The Facetime product allows records to be kept of all IM activity, important not only in SEC-regulated businesses, but also in government agencies which are obliged to keep copies and archives of all communications as public records. *Vericept* is another product with all-purpose monitoring capabilities, but focuses largely on its ability to monitor, block, and record IM activity. For organizations looking to not only block and monitor but also have a secure IM tool for internal communications, *Communicator Hub* is a proprietary IM tool that uses encryption to keep the contents of the IM secure from prying eyes.

General Monitoring (Sniffing at the Client) Tools

So far, the tools mentioned have largely been for an organization to monitor one particular type of personal Internet abuse. In general, these tools have been installed at a server on the network where it is able to snoop on traffic as it passes points in the network. However, there are a series of more powerful tools available for almost total user computer monitoring, classified under the heading of general monitoring tools. These tools are installed at the client and can create a record of almost everything a user does with the computer. Keystrokes are monitored and recorded, as are mouse movements. Snapshots of the screen can be taken frequently, in some cases as often as every second. The records can be written to a central database on a server, or they can even be e-mailed to another account.

These tools are often the subject of unsolicited commercial e-mail (a.k.a., spam) messages, targeted at people who suspect a spouse of infidelity or want to keep a close eye on children. However, these products are also targeted at businesses where complete records need to be kept before disciplining or terminating an employee, such as in cases where there is strong union support for worker rights.

Winwhatwhere is probably the original instantiation of this type of program. It has gone through several iterations and is marketed largely to businesses that need to do certain types of monitoring. There are a plethora of programs targeted at individuals suspecting family problems over the computer, including *Spy Agent* and *eBlaster*. Managers are often surprised at the sheer amount of data these programs provide. They can tell every program that is run, and for how long it was used. They can also tell every key that was pushed (and subsequently erased). This is often more information than a manager really wanted to know, so one should carefully consider the implications before implementing a general monitoring tool. Depending on the organization, employees may react strongly to such total monitoring, as discussed in the following section.

RECOMMENDED FIT BETWEEN GOALS AND MONITORING SOLUTIONS

After viewing an apparently massive list of potential solutions for electronic monitoring, one might wonder which product to choose and why. At this stage, one might be more concerned with fit between the goals for monitoring and the actual strategy to pursue. At the beginning of this chapter, three major goals for monitoring were listed: increasing productivity, bandwidth preservation, and legal liability reduction. A discussion below will detail the recommended fit for each of these goals.

If productivity is a major concern, one might begin with a passive but comprehensive logging tool. In this manner, productivity losses due to cyberslacking can be easily seen and measured, but it is not outwardly confrontational with employees. When an employee's productivity seems to fall, as observed by non-technical means, the data showing the amount of cyberslacking can be presented. This can be used as a means for implementing positive disciplinary measures, or for supporting a termination. In any event, when a situation occurs, and at periodic times throughout the year, employees should be reminded of the organization's policy about personal use of the Internet, and enforcement actions should be made clear. This is of course not to embarrass the potential offending party, but rather to remind the workers of the policy and show that it is enforced.

If legal liability is the major concern, a minimally intrusive means can also be used for the monitoring and recording of transmitted data. In December 2002, five major Wall Street brokerage houses were fined $1.65 million for not keeping e-mails the required two years. An organization can avoid these types of penalties by simply logging and maintaining records of Internet traffic without any review, except on an as required basis. The RIAA and other entertainment industry groups in early 2003 began warning organizations to actively ensure that their employees were not using company networks to access copyrighted music and video files, or that the companies themselves would be held liable. Often times, the RIAA has been supplying the IP addresses and times of access to companies and universities, identifying individuals who may have traded in music and video files, in effect doing the monitoring for the company. In the end, a company pursuing this strategy would be more concerned with record-keeping than record-reviewing.

An organization pursuing the third goal, bandwidth preservation, can likely use the least passive of all monitoring tools — simply observing bandwidth usage spikes where they occur and witnessing their relationship to organizational goals. A firm that sees the network pipe constantly full with apparently work-related material may want to investigate adding additional bandwidth resources. At the same time, organizations that suddenly block or "throttle" access to popular Internet destinations known for non-work-related information will likely solve many problems. Employees are more likely to realize that they need to get to work or identify another means of entertainment than they are to complain and cause a ruckus over no longer being able to access these sites at all or at high speeds.

RECOMMENDED FIT BETWEEN PRODUCTS, GOALS, AND CORPORATE CULTURE

This chapter has detailed several types of control mechanisms. It has listed goals for monitoring, applications to monitor, and means of accomplishing the monitoring. Furthermore, it lists names of actual tools that can be used to accomplish the monitoring. What this chapter so far has not discussed is whether or not the monitoring will actually work, and if it will produce unintended consequences and side effects.

In a series of studies done in the late 1990s, it was found that the presence of monitoring indeed has a significant positive effect on keeping employees on task (Urbaczewski & Jessup, 2001). There are also several anecdotes in the press that would confirm this finding in different organizations. However, it was also found that monitored employees were more likely to turn over and less likely to participate in other organizational activities. Could this happen in all organizations? Possibly, but the key to remember when establishing monitoring is:

"Make the monitoring strategy fit the corporate culture."

Some organizations have a culture where monitoring of most (if not all) activities is simply expected. These industries generally include those with a large amount of cash or financially related transactions (e.g., banks, casinos) or that deal with physical and/or national security (CIA, FBI, R&D labs, etc.). In these cases, monitoring fits in perfectly with the culture, and if the organization is already monitoring employees in many other ways, it would make sense to add monitoring capabilities to computer systems (if employees even have access to them).

Other organizations have cultures which generally do not practice monitoring. Some industries, like publishing, academia, and other "civil liberties" organizations, do not generally monitor their employees, as the foundation for the organizations is centered around the freedom of speech and the unfettered search for the truth. The introduction of monitoring in these industries will likely result in a culture class between employees and management (Simmers, 2002).

The question then becomes, "How does one then reap the benefits of monitoring without angering employees and creating unwanted stress in the organization?" This question is best left to other chapters in this text, but the general idea in this context is communication of the control mechanisms to the employees, with a clear understanding of how the mechanisms support corporate goals and principles. Explicit statements of who will be monitored, what will be monitored, and when monitoring will occur should also be communicated to the employees, largely through acceptable use policies.

REFERENCES

Fonseca, B. Anna Kournikova worm hits the United States. *Infoworld Online.* Retrieved on June 5, 2003 from: http://www.infoworld.com/articles/hn/xml/01/02/12/010212hnanna.xml.

McCullagh, D. Scarfo: Feds plead for secrecy. *Wired Online.* Retrieved on June 5, 2003 from: http://www.wired.com/news/politics/0,1283,46329,00.html.

Reuters (author unknown). New sites top spots for work surfing. *CNN.com.* Retrieved on June 5, 2003 from: http://www.cnn.com/2002/TECH/internet/09/23/workplace.surfing.reut/index.html.

Simmers, C.A. (2002). Aligning Internet usage with business priorities. *Communications of the ACM*, (January), 71-74.

Swanson, S. (2002). Employers take a closer look. *Information Week*, (July 15), 40-41.

Urbaczewski, A. & Jessup, L.M. (2002). Does electronic monitoring of employee Internet usage work? Communications *of the ACM*, (January), 80-83.

Chapter IX

Convergence or Divergence? Web Usage in the Workplace in Nigeria, Malaysia, and the United States

Claire A. Simmers
Saint Joseph's University, USA

Murugan Anandarajan
Drexel University, USA

ABSTRACT

This study sets out to examine whether employee web usage patterns, attitudes toward web usage in the workplace, and organizational policies are more similar (convergence thesis) or less similar (divergence thesis) in three countries: Nigeria (n = 224), Malaysia (n = 107), and the United States (n = 334). Our results show general support for the divergence thesis. We found strong differences in employee usage patterns by country, even after controlling for differences in several demographic variables. However, there is less support for the divergence thesis in attitudes and organizational policies. In half of the eight indicators of employee attitudes, there were no differences among the three countries. Agreement that personal web usage at work is acceptable behavior is widespread. Other common perceptions are that companies tolerate personal web searches and that Internet usage policies are not enforced.

INTRODUCTION

Cross-cultural researchers and practitioners concur that there is a need to better understand and manage the tension between the durability of national cultures (divergence) and the closer, more frequent interactions among nations (convergence) (Adler, 1997; Brodbeck et al., 2000). National boundaries are increasingly permeable; the number of multinational corporations is increasing, many people are employed transnationally (Hodgetts et al., 1999) and participation in cross-cultural teams is commonplace. There is an increasingly complex matrix of global interaction points in the workplace made possible by communication innovations, particularly in information technology and the widespread usage of the World Wide Web. At the same time, there is attention to national pride, thus the tension between convergence and divergence heightens. Few would argue that in the last decade of the 20th century, the World Wide Web revolutionized the way we work. The business world has been "blown to bits" (Evans & Wurster, 2000), digitized (Cronin, 2000), globalized (Ohmae, 2000) and uniquely challenged (Drucker, 1999). Consideration of the issues raised by these unprecedented changes invites the exploration of an important question, specifically, the relationship between national cultural diversity and managing human resources in a digital economy. For instance, given the escalating importance of the web in the workplace, the more we know about workplace information technology (IT) behaviors and attitudes, particularly workplace usage of the World Wide Web, the more effectively and efficiently we can manage. Empirical data on cultural variation in web usage and attitudes can be helpful for those who deal with employees in the Internet-anchored workplace, particularly those in human resources and information technology.

In this information/knowledge economy, people are critical sources of sustainable competitive advantage (Delaney & Huselid, 1996; Wright et al., 1994). Resource and knowledge based theories of the firm suggest that organizational survival and success depend upon how well human resources are deployed and managed (Davis, 1995; Erez, 1994; Triandis, 1994). The importance of effective management of human capital rather than physical capital as the ultimate determinant of organizational performance is often emphasized (Youndt et al., 1996). An important aspect of managing human capital in the 21st century workplace is managing the interface between humans and information technology—particularly the Internet. Many have argued that web usage at work is being misused and that there is a high cost in giving web access to employees (Naughton, 1999). Others counter that employees need to be given access to the web in order to enhance their skills and enhance competitive advantage (Kerwin et al., 2000). Research insights for mangers on the relationship between national culture and employee web usage and attitude will facilitate the development and enforcement of policies on usage and monitoring of the Internet. If web usage and attitudes differ as a function of national culture, then information technology training, monitoring policies, and system implementations need to consider national culture as an important moderating variable. In other words, the more web usage and attitudes differ by national culture, the more need for web policies that take into account heterogeneous cultural environments (Dirksen, 2000).

Few studies examine employee practices and attitudes about web usage across cultures. In this chapter, we use a national culture approach to frame our investigation into employee workplace web usage and attitudes in three countries: Nigeria, Malaysia, and the United States. This framework is consistent with the thinking and research of a number of researchers including Hofstede (1993), Newman and Nollen (1996), Smith, Dugan and Trompenaars (1996), and Trompenaars (1993). Specifically, we wanted to know if there were national differences in employees' responses on: (a) self-reported frequency of accessing web pages at work; (b) perceived attitudes on personal web usage at work; and (c) organizational policies on controlling workplace web usage. *Similarities* in responses across nations would lend support for the convergence theory while *differences* across nations would lend support for the divergence theory. Our findings can foster the development of culturally sensitive information technology training, usage policies, and monitoring procedures, as well as facilitate productive Internet usage.

THEORY

Increased international business activity and emphasis on globalization have rekindled interests in the convergence-divergence theory, which dominated much of U.S. and European management research in the 1950s and 1960s (Dowling, 1999). The convergence theory states that national cultures are slowly becoming more homogenized (reflecting a shrinking world). This is a result of the global economy, information technology, and similar educational and work experiences (Adler, 1983; Child 1981). Given the thesis that increasing global interconnectivity and interdependence follows a global market economy (Wright & Ricks, 1994), it does seem reasonable to expect that there will be increasing cultural similarity in thinking and values. The convergence thesis maintains that economic ideology drives values. As a result, industrialized nations will share common values with regards to economic activity and work-related behavior (England & Lee, 1974). Convergence implies that as developing countries industrialize and embrace free-market capitalism and technology, then they will adopt the ideological values of the developed industrialized world (Kelley et al., 1995; Priem et al., 2000). Advocates of the convergence theory hold that employee workplace web usage and attitudes — irrespective of culture — will, over time, tend toward commonality and that these commonalities are present in all industrial or industrializing societies (Ralston et al., 1993). Although convergence is often equated with Westernization or Americanization, U.S. values appear to be affected and American value systems are becoming less nationally based (Fernandez et al., 1997).

The divergence perspective recognizes country and cultural differences. The main hypothesis is that national culture continues to be a dominating influence on individuals' attitudes and behaviors (Hofstede, 1997). The proponents maintain that culture is deep-rooted and drives values of any society beyond capitalism or economic ideology. They expect the value systems of people in the workforce to remain largely unchanged even if they adopt and have widespread web usage (Ricks et al., 1990; Ralston et al., 1995). Moreover, the proponents believe that national or regional cultural influences will continue to value diversity among even fully industrialized societies. Hence, the divergent perspective is consistent with the dominant perspective of some cross-cultural theorists (e.g., Hofstede, 1980, 1997; Adler, 1997) who emphasize that all management practices are culturally determined. Cross-cultural research is well established and has cataloged how basic assumptions, values, and behavioral norms vary across cultures (Hampden-

Turner & Trompenaars, 1993; Hofstede, 1980; Schwartz, 1992; Triandis, 1989). Hofstede (1997) argues that although individuals in organizations may appear to be more similar, this similarity is the result of the organizational acculturation process, not the convergence of national cultures.

Web Usage and National Culture

Within a global competitive environment, web usage and attitudes about web usage in the workplace take on new meanings and directions and there are important implications for top management and for information system (IS) units in every institution. In this chapter we define web usage as accessing different types of web pages (Anandarajan et al., 2000). Administrating web usage in today's changing workplace is a challenge and the line between productive and non-productive web usage is getting fuzzier (Sunoo, 1996). Increasingly, IS units are called upon to monitor and control web usage while upper level decision makers see the web as a competitive tool. While growing, research on web usage in the U.S. is still sparse and there are few cross-cultural comparisons (Montealegre, 1998). If the power of the web is to be harnessed for competitive advantage, IS and top management need to better monitor and control web usage, while facilitating and encouraging productive web usage. Furthermore, they need to better understand the national culture dimension of IT.

Using the Internet can create many desirable organizational outcomes — lowering the cost of communication, restructuring how work is done, supply chain management, and improving business practices and integration. However, using the Internet can also generate undesirable outcomes — loss of intellectual property, sexual harassment lawsuits, productivity losses due to surfing abuse, security threats, and network bandwidth overload by visiting websites for travel, leisure, sports, and news, for example. The link between usage of the web and national culture is not clear and there is a lack of research on national culture as an explanation of either positive or negative web usage in the workplace. This is surprising since cultural values have been shown to have a significant impact on a wide array of business practices such as compensation (Schuler & Nogovsky, 1998), leadership (Brodbeck et al., 2000), global research and development activities (Jones & Teegen, 2001), and software piracy (Husted, 2000).

Technical, social, and cultural reference frames co-mingle in an information technological infrastructure. Most information technology research looks at organizational or corporate culture and individual reasons for web usage (Davis, Bagozzi, & Warshaw, 1989) and seldom considers the impact of national culture (Dirksen, 2000). Mansell and Wehn (1998) suggest that many common assumptions rooted in the U.S. about information technology usage patterns may not be similar in other national cultures. Consequently, drawing on the convergence-divergence theory discussed earlier, similarity in patterns of web usage will lend support for the convergence theory and differences in patterns of web usage will lend support for the divergence theory, thus leading us to hypothesize:

H1: Patterns of web usage will be more similar than different among the countries.

Attitudes Toward Personal Web Usage at Work and Organizational Controls and National Culture

A model of cross-cultural ethics would posit that attitudes would vary by national culture (Cohen, Pant, & Sharp, 1996; Husted, 2000; Vitell, Nwachukwu, & Barnes, 1993). Cross-cultural ethics posits that decisions involving such ethical situations as piracy and questionable accounting will be influenced by values (Husted, 2000). Conversely, because of the global economy and the influence of information technology, the convergence theory would lead us to expect that there would be few differences in attitudes about using the web for personal searches while at work. There is a common language of bytes, random access memory (RAM), firewalls, and direct service lines (DSL) that transcends national boundaries. People using information technology in general and the web in particular, may adopt similar patterns of attitudes transcending their national culture differences (Ohmae, 1999). The convergence theory would suggest that people are becoming more similar in their attitudes on personal web usage. Additionally, as organizations become increasingly global, they will standardize procedures and policies, especially in information technology, with security protocols and usage reports. Hence, we hypothesize:

H2a: Attitudes about personal web usage will be more similar than different among the countries.

H2b: Organizational policies on web usage will be more similar than different among the countries.

Thus, given the preceding arguments, we have framed our hypotheses to indicate our preference for the convergence perspective, which posits the convergence of behaviors (web usage) and value dimensions (attitudes about personal web usage at work) with increasing industrialization and globalization. We do recognize that there is a lag in the chain of change and that there are value dimensions that remain largely divergent. However, we need to continue empirical investigation to show support for the logic of our position and we can discount neither the convergent nor the divergent perspectives without empirical study.

METHODS

Research Setting

We chose countries for our research setting that represented geographical, economic, technological, and national culture variances. Brodbeck et al. (2000) have shown that cultural variance is higher in samples with countries from different geopolitical regions. More importantly, our choices reflected a gap in research related to the adoption and usage of the web in less developed countries (LDC) (Avgerou, 1996). This lack of research is partially related to the fact that until the early part of the 1990s, the diffusion of information technology (IT) in many regions such as Africa, Asia, and Latin America was extremely low (Rigg & Goodman, 1992; Odedra, Lawrie, Bennett, & Goodman, 1993). However, the LDCs recognize the importance of information systems (Ehikhamenor, 1999) and microcomputer purchases in the business sector of these regions are growing at an annual rate of 90% (Plunkett's InfoTech Industry Almanac, 1997).

Nigeria

Nigeria, although an LDC, is one of the largest economies in the Sub-Sahara region of Africa (Feldman, 1992) and many major multinational corporations and their affiliates conduct business there (Jason, 1997; Thompson, 1994). In Nigeria, the Gross Domestic Product (GDP) is as follows: purchasing power parity is $110.5 billion (1999 est.), the per capita purchasing power parity is $970 (1999 est.), and in 1999 the number of Internet Service Providers (ISPs) is five (CIA 2000 World Factbook). Although Nigeria is a diverse society with approximately 300 ethnic and sub-ethnic groups with as many distinct languages and dialects, the family culture value system is evenly applicable to most of Nigerian society regardless of ethnic affiliation (Gannon, 1994).

Malaysia

Poverty rates have fallen dramatically over the past 20 years in this former British colony of 20 million people. It has a fast growing economy, ranking it as a leading LDC. In Malaysia, the GDP is as follows: purchasing power parity is $229.1 billion (1999 est.), the per capita purchasing power parity is $10,700 (1999 est.), and the number of Internet Service Providers (ISPs) is eight (1999) (CIA 2000 World Factbook). The Chinese, Malays, and Indians are the major cultural segments in Malaysia. Government efforts to build national unity and identity, such as the increasing use of Malay language in public life, has met with some success, although fundamental differences in culture have been found to exist in negotiation styles (Loo, 2000). We follow Lim and Baron (1997) in using Malaysia as a national entity.

United States

The U.S. is the largest economy as evidenced by GDP as follows: purchasing power parity of $9.255 trillion (1999 est.), a per capita purchasing power parity of $33,900 (1999 est.), and 7,600 (1999 est.) Internet Service Providers (CIA 2000 World Factbook).

Data Collection and Sample Profile

The results reported in this chapter are part of a larger study on Internet usage in the workplace. The relevant questions can be found in Appendix A. The survey was piloted tested and revisions made on this basis (Anandarajan et al., 2000). The data was collected from a convenience sample drawn from working adult populations in all three countries.

Due to unreliable postal services, the need to establish personal relationships, and the lack of computers in the general population, data was collected differently in Nigeria and Malaysia. Similar to data collection methods used by Steensma, Marino, Weaver and Dickson (2000) an onsite structured questionnaire collection process was used in both of these countries. Trained interviewers scheduled appointments, presented the key contact with the surveys, answered any questions, and returned to collect the completed questionnaires. A similar method was employed in Malaysia. In the U.S., because of higher computer usage, a reliable mail system, and general tendency to respond to "cold-call" surveys, a survey was mailed to a randomly selected sample of 3,000 from the alumni database of a Northeastern university.

A total of 794 usable questionnaires were returned (Nigeria — 237; Malaysia — 113; and the U.S. — 444). Only those respondents using the Internet at work were examined in this study. The total was 665, with the following breakdown — 224 from Nigeria, 107 from Malaysia, and 334 from the U.S.

Profile of Internet Users

Table 1 shows the demographic statistics for the sample.

Two/thirds of the Nigerian and U.S. samples were men, while the Malaysian sample was evenly divided. The Nigerian and Malaysian respondents were considerably younger than those from the U.S. In Nigeria, 72.6% of the sample reported income of less than $20,000; the average salary range for the Malaysian sample was between $20,001 and $30,000; and in the U.S., it was between $45,001 to $65,000. More than 50% of the respondents worked at businesses with fewer than 1,000 employees. The respondents in Nigeria were evenly spread among the different professional levels. More of the Malaysian and U.S. respondents (39% each) were professionals than in Nigeria (22%). The Malaysian and the U.S. respondents reported more Internet usage outside of work than the Nigerian respondents did. The respondents in all three nations confirmed that their companies had an Internet presence by reporting that their companies had a website.

There were a variety of industries represented in the sample. In Nigeria, three quarters of the respondents worked in the services sector or the finance, insurance or real estate sector. Half of the Malaysian respondents reported working in the services sector. United States respondents worked in a cross-section of industries.

Measures

Independent Variables

Surveys were assigned a country code — Nigeria = 1, U.S. = 2, Malaysia = 3 — establishing three groups. There were eight demographic variables.

Table 1. Background Demographics

	Nigeria	U.S.	Malaysia	Total
Total responses	237	444	113	794
No access at work	13 (5.5%)	110 (24.8%)	6 (5.3%)	129 (16.2%)
Access at work	224 (94.5%)	334 (75.2%)	107 (94.7%)	665 (83.8%)
Gender	217	334	103	654
Male	145 (66.8%)	212 (63.5%)	57 (55.3%)	414 (63.3%)
Female	72 (33.2%)	122 (36.5%)	46 (44.7%)	240 (36.7%)
Type of Business				
Manufacturing	4%	16%	11%	11%
Services	41%	20%	50%	32%
Wholesale, Retail Trade	5%	2%	1%	3%
Finance, Insurance, Real Estate	30%	14%	0%	17%
Education	2%	12%	7%	8%
Government	4%	11%	11%	8%
Self-Employed	0%	3%	0%	2%
Other	14%	22%	20%	19%
Size of Business				
1-999 employees	151 (68.4%)	162 (48.8%)	57 (53.8%)	370 (56.2%)
1,000-9,999 employees	39 (17.6%)	86 (25.9%)	30 (28.3%)	155 (23.5%)
more than 10,000 employees	31 (14%)	84 (25.3%)	19(17.9%)	134 (20.3%)
Current Position				
Top Level Manager	25 (11.7%)	57 (17.1%)	5 (4.9%)	87 (13.4%)
Middle Level Manager	46 (21.6%)	66 (19.8%)	8 (7.8%)	120 (18.5%)
Lower Level Manager	40 (18.8%)	30 (9.0%)	10 (12.5%)	80 (12.3%)
Professional	48 (22.5%)	130 (39.0%)	40 (39.2%)	218 (33.6%)
Administrative Support	37 (17.4%)	21 (6.3%)	20 (19.6%)	78 (12.0%)
Other	17 (8.0%)	29 (8.7%)	19 (18.6%)	65 (10.0%)
Age				
20-30 years	128 (58.4%)	68 (20.7%)	66 (64.1%)	262 (40.3%)
31-40 years	63 (28.8%)	104 (31.7%)	32 (31.1)	199 (30.6%)
41-50 years	24 (11.0%)	88 (26.8%)	2 (1.9%)	114 (17.5%)
51-60 years	3 (0.4%)	51 (15.5%)	3 (2.9%)	57 (8.8%)
more than 60 years	1 (0.5%)	16 (4.9%)	0	17 (2.6%)
Web Usage Outside of Work				
Yes	98 (43.8%)	253 (75.7%)	82 (78.1%)	433 (65.3%)

Business or industry was measured by eight categories and size of the company was measured by number of employees from "1" representing 1-49 to "8" representing more than 10,000. Due to insufficient numbers in each category for each country, the categories were collapsed from eight to three with small companies represented with "1" (1-999), medium companies represented with "2" (1,000-9,999), and large companies represented with "3" (greater than 10,000). Respondents were asked to describe their current position as top level manager, middle level manager, lower level manager, professional, administrative support, and other. Salary options ranged from "1" representing less than $20,000 to "7" representing more than $120,000. Age was reported in year and then coded to represent ranges. Gender was coded "1" for male and "2" for female. Having a company website and accessing the Internet were coded "1" for yes and "2" for no.

Dependent Variables

We included three sets of areas to test for potential similarities or differences in web behavior and attitudes: employee Internet usage, attitudes on Internet usage, and information on organizational policies on monitoring Internet usage.

To measure employee web usage we used types of web pages accessed (Cronin, 1995). Each respondent was asked to indicate how likely it was that s/he would access 10 different kinds of web pages while at work — "1" = very unlikely to "5" = very likely. Examples included competitor websites, arts and entertainment websites, customer websites, and sports/news websites.

Attitudes on Internet usage were assessed by asking respondents to give their opinion of uses of the Internet while at work by answering three questions — "1" = strongly disagree to "5" strongly disagree. The three questions were: "I feel that using the Internet for personal searches is acceptable," "In my company, it seems that accessing the Internet for personal searches is tolerated," and "I feel my company should block access to Internet sites which are deemed inappropriate for business use."

Five items gathered information about organizational policies on Internet usage. The first item asked for respondents to indicate on a scale of "1" = strongly disagree to "5" strongly disagree, if his/her company considers it important to provide its employees with regular reports on Internet usage. The other four items, each tapped by a questionnaire item measured as a "Yes" (coded 1) or "No" (coded 2) were: "Does your company block access to certain Internet sites?," "Do you have additional passwords to access the Internet?," "My company has clearly stated Internet usage policies," and "My company strictly enforces its Internet policy."

Data Analysis

The general linear model multivariate procedure used a technique to measure analysis of variance for multiple dependent variables by multiple factor variables. This procedure allows for the testing of unbalanced designs (different number of cases in each cell). The first step was to use analysis of variance to test for demographic differences that might influence the responses to the dependent variables. We then examined the general relationships among the variables by running a general linear model testing for significant relationships among multiple independent and dependent variables. We sought evidence of similarities or differences among the countries on the dependent measures with the significant demographic variables as controls. We also used post hoc comparisons to identify which nations were significantly different from each other if a significant F ratio for the entire model was obtained. We used the conservative Scheffe's test of significance post hoc tests. The significance level of the Scheffe test is designed to allow all possible linear combinations of group means to be tested, requiring a larger difference between means for significance (Huck, Cormier, & Bounds, 1974).

RESULTS

Internet User Demographics

Analysis of variance using each of the eight demographic variables as the dependent variable and country as the independent variable resulted in significant differences among countries in six of the variables: (1) business and industry ($F = 7.315, p < .001$); (2) size ($F = 11.575, p < .000$); (3) position ($F = 15.854, p < .000$); (4) salary ($F = 316.946, p < .000$); (5) age ($F = 62.534, p < .000$); and (6) use of the Internet outside of work ($F = 38.704, p < .000$). Because of this, we entered these demographic variables as control variables.

Divergence or Convergence

The means and standard deviations for the dependent variables (employee web usage, attitudes and information) are given in Table 2.

Table 2. Means and Standard Deviations of Dependent Variables

	Nigeria		U.S.		Malaysia	
Web Usage [a]	Mean	Standard Deviation	Mean	Standard Deviation	Mean	Standard Deviation
Competitor	3.24	1.20	3.06	1.49	2.99	1.27
Government/Research	3.13	0.98	3.35	1.32	3.57	1.19
General Interest	3.84	2.16	3.03	1.10	3.65	1.03
Suppliers	3.40	1.05	2.72	1.30	3.44	1.10
Customers	3.31	1.06	2.67	1.46	3.02	1.27
Arts and Entertainment	3.08	1.08	2.41	1.23	3.15	1.11
Travel and Leisure	2.96	1.05	2.55	1.31	2.98	1.16
Living/Consumer	3.14	1.15	2.31	1.21	2.87	1.14
Business and Financial	3.79	1.04	3.35	1.32	3.11	1.12
Sports/News	3.73	1.10	2.60	1.39	3.14	1.19
Attitudes [b]						
I think personal web searches at work are acceptable	3.77	0.91	3.53	1.11	3.68	1.05
My company tolerates personal web searches	3.52	1.03	3.51	1.02	3.60	0.92
My company should block access to certain web pages	3.68	1.27	2.78	1.26	2.91	1.20
My company considers regular web usage reports important	2.99	1.13	2.03	1.01	3.07	1.05
Organizational Policies [c]						
	Yes	No	Yes	No	Yes	No
My company blocks access to certain web pages	62 (28.3%)	157 (71.7%)	59 (19.8%)	239 (80.2%)	21 (21.4%)	77 (78.6%)
My company has additional passwords for web access	130 (59.6%)	88 (40.4%)	103 (34.0%)	200 (66.0%)	27 (26.7%)	74 (73.3%)
My company has clearly stated Internet usage policies	131 (60.1%)	87 (39.9%)	150 (49.5%)	153 (50.5%)	40 (40.0%)	60 (60.0%)
My company strictly enforces its Internet policy	63 (29.0%)	154 (71.0%)	90 (30.5%)	205 (69.5%)	27 (27.3%)	72 (72.7%)

[a] *The question is: how likely are you to access the following web pages while at work. Scale is: 1 = very unlikely; 2 = unlikely; 3 = likely; 4 = most likely; 5 = very likely*

[b] *The question is to agree or disagree with subsequent statements. Scale is: 1 = strongly disagree; 2 = disagree; 3 = neither agree nor disagree; 4 = agree; 5 = strongly agree*

[c] *Scale is 1 = yes; 2 = no*

Table 3. Differences in Employee Web Usage — Accessing Types of Websites

	Competitors *F* / Sig.	Government/ Research *F* / Sig.	General Interest *F* / Sig.	Suppliers *F* / Sig.	Customers *F* / Sig.	Arts/ Entertainment *F* / Sig.	Travel/ Leisure *F* / Sig.	Living/ Consumer *F* / Sig.	Business/ Financial *F* / Sig.	Sports/ News *F* / Sig.
Overall Model	3.681 .000 Adj R^2 .142	3.008 .000 Adj R^2 .110	3.188 .000 Adj R^2 .119	3.129 .000 Adj R^2 .116	3.503 .000 Adj R^2 .134	2.552 .000 Adj R^2 .088	1.593 .020 Adj R^2 .035	3.629 .000 Adj R^2 .140	3.428 .000 Adj R^2 .131	3.841 .000 Adj R^2 .149
Type of Business	2.119 .040	7.065 .000	n/s	n/s	3.890 .000	n/s	n/s	n/s	4.473 .000	n/s
Size of Business	n/s	2.221 .025	n/s	n/s	n/s	n/s	n/s	n/s	n/s	n/s
Position	5.165 .000	n/s	n/s	3.137 .008	4.946 .000	n/s	n/s	2.412 .035	2.663 .022	n/s
Salary	2.926 .008	n/s	n/s	n/s	n/s	n/s	n/s	n/s	n/s	n/s
Age	4.181 .001	n/s	2.467 .032	n/s	n/s	n/s	n/s	n/s	n/s	2.202 .053
Website use outside of work	9.458 .002	n/s	3.718 .054	7.893 .005	n/s	n/s	n/s	n/s	6.209 .013	n/s
Country	n/s	4.852 .008	18.071 .000	14.841 .000	3.460 .032	8.749 .000	6.791 .001	16.140 .000	7.885 .000	18.176 .000

F / Sig. = F value and significance level
n/s = not significant
Adj R^2 *= Adjusted* R^2

Employee Web Usage

The multivariate analysis of variance for the measures of employee web usage was found to be significantly different among Nigeria, Malaysia, and the U.S. ($F = 6.577, p < .000$) by the Wilks' Lambda criterion. Tests of Between-Subjects Effects showed significant differences among the three countries in accessing nine of the 10 types of web pages. Results are given in Table 3.

The results of the post hoc investigation are shown in Table 4. Respondents in the U.S., on average, are significantly less likely to access five of the nine types of web pages (general interest, suppliers, arts/entertainment, travel/leisure, and living/consumer) than those respondents from either Nigeria or Malaysia. Malaysians are less likely to access competitor web pages than either Nigerians or those from the U.S. Nigerians are more likely to access business and financial web pages while at work than the respondents from the other two countries. This usage pattern might be linked to the lower web access outside of work reported by Nigerians. Of particular interest are the results on accessing sports/news websites while at work. All three countries report

Table 4. Scheffe's Test of Multiple Comparisons for Web Usage

Websites accessed with significant differences:	*(I) Nation*	*(J) Nation*	**Mean Difference (I-J)**	**Sig.**
Government/ Research	Nigeria	Malaysia	-.4652	.010
General Interest	Nigeria	U.S.	.6740	.000
	U.S.	Malaysia	-.7012	.000
Suppliers	Nigeria	U.S.	.6650	.000
	U.S.	Malaysia	-.7683	.000
Customers	Nigeria	U.S.	.5710	.000
Arts/ Entertainment	Nigeria	U.S.	.6591	.000
	U.S.	Malaysia	-.5544	.001
Travel/Leisure	Nigeria	U.S.	.4141	.001
	U.S.	Malaysia	-.4364	.016
Living/Consumer	Nigeria	U.S.	.8914	.000
	U.S.	Malaysia	-.5544	.001
Business/ Financial	Nigeria	U.S.	.4898	.000
	Nigeria	Malaysia	.7446	.000
Sports/News	Nigeria	U.S.	1.1437	.000
	Nigeria	Malaysia	.6750	.000
	U.S.	Malaysia	-.4687	.015

significantly different usage patterns, with the Nigerians most likely to access these pages while at work (mean = 3.73), the Malaysians likely (mean = 3.14), and those from the U.S. unlikely (mean = 2.60). In summary, employee web usage patterns are largely different among the three countries, thus Hypothesis 1 is not supported.

Attitudes Toward Personal Usage and Information on Organizational Web Usage Policies

The multivariate analysis of variance for the measures of employee attitudes and organizational web usage policies was found to be significantly different among Nigeria, Malaysia, and the U.S. ($F = 6.713, p < .000$) by the Wilks' Lambda criterion. Tests of Between-Subjects Effects showed significant differences among the three countries in four of the eight attitudes and web usage policies at work. There were significant differences in attitudes about companies blocking access to Internet sites, on the importance that companies place on providing regular Internet usage reports, on additional passwords to access the Internet, and on whether companies have clearly stated Internet usage policies. Results are given in Table 5.

The results of the post hoc investigation are shown in Table 6. Nigerians agree that companies should block access to certain web pages — an attitude that is not shared by either the U.S. respondents or the Malaysian respondents. Nigerians also report that their companies have additional passwords to access the Internet, which is not reported in either Malaysia or the U.S. Malaysians report that they have clearly stated Internet policies. This is significantly different from the Nigerian respondents. In summary, employees' attitudes and information on organizational Internet policies are different among the three countries — thus neither Hypothesis 2a nor 2b is supported.

DISCUSSION

Our results present general support for the divergence thesis. There are clear differences in employee usage patterns by country, even after controlling for differences in several demographic variables; however, there are fewer differences in attitudes and organizational policies. In half of these indicators, there were no differences among the three countries. Particularly important was the general agreement that personal web searches at work are acceptable

Table 5. Differences in Attitudes on Personal Usage and Opinions on Company Controls of Personal Usage

	Personal Usage is Acceptable	Personal Usage is Tolerated	Should Block Websites	Issues Usage Reports	Does Block Websites	Has Passwords	Has Usage Policies	Enforces Policies
	F / Sig.	*F* / Sig.	*F* / Sig.	*F* / Sig.	*F* / Sig.	*F* / Sig.	*F* / Sig.	*F* / Sig.
Overall Model	2.825 .000 Adj R^2 .103	n/s	3.077 .000 Adj R^2 .116	4.414 .000 Adj R^2 .177	1.617 .017 Adj R^2 .037	2.721 .000 Adj R^2 .098	3.336 .000 Adj R^2 .128	n/s
Type of Business	2.777 .008	2.682 .010	n/s	n/s	2.573 .013	n/s	n/s	n/s
Size of Business	n/s	n/s	n/s	n/s	n/s	n/s	6.783 .000	2.549 .010
Position	n/s	n/s	n/s	n/s	3.473 .004	n/s	2.708 .020	n/s
Salary	n/s	n/s	n/s	n/s	2.141 .047	n/s	n/s	n/s
Age	n/s	n/s	n/s	n/s	n/s	2.731 .019	n/s	n/s
Website Use Outside of Work	n/s	n/s	11.260 .001	n/s	n/s	n/s	n/s	n/s
Country	n/s	n/s	12.878 .000	17.718 .000	n/s	15.936 .000	4.795 .009	n/s

F/Sig. = F value and significance level
n/s = not significant
Adj R^2 = Adjusted R^2

Table 6. Scheffe's Test of Multiple Comparisons for Attitudes and Perceptions

Dependent Variables with Significant Differences	*(I) Country*	*(J) Country*	**Mean Difference (I-J)**	**Std. Error**	**Sig.**
Should Block	Nigeria	U.S.	.8955	.1184	.000
	Nigeria	Malaysia	.7610	.1649	.000
Has Usage Reports	Nigeria	U.S.	.9562	.1013	.000
	U.S.	Malaysia	-1.0402	.1339	.000
Has Passwords	Nigeria	U.S.	-.2757	4.463E-02	.000
	Nigeria	Malaysia	-.3249	6.218E-02	.000
Has Policies	Nigeria	Malaysia	-.2044	6.196E-02	.005

and that the companies tolerate personal searches while at work. Taken together, what do our findings say about the impact of national culture on employee web usage and attitudes?

First, our findings should be interpreted in the context of a rapidly changing environment. The usage reported in this study is modest and Internet usage in Nigeria and Malaysia is still in its infancy. The uncertainty and newness of the Internet may explain some of the responses. While Nigerian and Malaysian respondents agree that using the Internet for personal searches is acceptable, the respondents from the U.S. are more ambivalent. None of the respondents have strong opinions on whether their companies should block access to inappropriate websites. Most of the respondents in the three countries thought that their companies tolerated personal searches and most questioned whether their companies considered regular reports on Internet usage important. Perceptions of organizational policies on monitoring and security methods adopted in the work place indicate a lack of consistency in organizational policies. Not blocking access to selective websites was reported by at least three-quarters of the respondents. Overwhelmingly, the respondents report that their companies do not strictly enforce Internet policies. In Nigeria, approximately 60% of the respondents reported additional passwords were required and Nigerian respondents thought that Internet policies were clearly stated. However, in the U.S. and Malaysia, security and monitoring were less stringent than in Nigeria, with more than 60% reporting that additional passwords were not required. Clearly stated Internet policies were far less common in the U.S. and Malaysia than in Nigeria.

We believe our findings indicate that the need to find a balance between excessive control and excessive freedom will be a continuing issue with country specific implementation considerations important for success. If the anticipated increase in web usage in the global economy occurs, it is probable that IS solutions need to emphasize a behavior modification orientation rather than an access control orientation. Attitudes and perceptions were remarkably similar across the three nations studied and suggest support for the developing presence of a more homogeneous global outlook on information technology policies and procedures.

The use of a convenience sample of only three nations is a major limitation in this study. Level of economic development has only been indirectly controlled by using salary and position as control variables in the data analyses. The generalizability of our results awaits additional empirical work. The cross sectional nature of our study also is a limitation and common method bias cannot be ruled out.

However, we feel that we have started an important line of inquiry. Web usage is growing and those organizations that are able to creatively use it to more effectively manage costs and to better satisfy customers will be at a competitive advantage. The increasing significance of the web to the organization is being seen throughout the global marketplace. The results of this work may seem most important to IS units because they are generally tasked with the responsibility of setting up and implementing IT control systems. However, the results also offer possible meaning for those in human resource management and for top organizational decision-makers as national culture appears to continue to be an important influence in the increasingly Internet-anchored workplace.

REFERENCES

Adler, N.J. (1983). Cross-cultural management research: The ostrich and the trend. *Academy of Management Review*, 8, 226-232.

Adler, N.J. (1997). *International Dimensions of Organizational Behavior* (3rd ed.). Cincinnati, OH: South-Western College Publishing.

Anandarajan, M., Simmers, C.A., & Igbaria, M. (2000). An exploratory investigation of the antecedents and impact of Internet usage: An individual perspective. *Behavior and Information Technology*, 19(1), 69-85.

Avgerou, C. (1996). How can information technology enable developing countries to integrate into the global economy? In P. Palvia, S. Palvia, & E. Roche (Eds.), *Information Technology and Systems Management*. Westford, MA: Ivy League Publishing.

Brodbeck, F.C., Frese, M., Akerblom, S., Audia, G., Bakacsi, G., Bendova, H., Bodega, et al. (2000). Cultural variation of leadership prototypes across 22 European countries. *Journal of Occupational and Organizational Psychology,* 73, 1-29.

The Central Intelligence Agency (CIA) 2000 World Factbook (2000). Retrieved January 4, 2001 from the World Wide Web: http://www.cia.gov/cia/publications/factbook/index.html.

Child, J.D. (1981). Culture, contingency and capitalism in the cross-national study of organizations. In L.L. Cummings & B.M. Staw (Eds.), *Research in Organizational Behavior,* 3, 303-356. Greenwich, CT: JAI Press.

Cohen, J.R., Pant, L.W., & Sharp, D.J. (1996). A methodological note on cross-cultural accounting ethics research. *International Journal of Accounting*, 31, 55-66.

Cronin, M. (1995). *Doing More Business on the Internet* (2nd ed.). New York: International Thompson Publishing Inc.

Cronin, M. (2000). *Unchained Value*. Boston, MA: Harvard Business School Press.

Davis, D. D. (1995). Form, function, and strategy in boundariless organizations. In A. Howard (Ed.), *The Changing Nature of Work.* San Francisco, CA: Jossey Bass.

Davis, F.D., Bagozzi, R.P., & Warshaw, P.R. (1989). User acceptance of computer technology: A comparison of two theoretical models. *Management Science*, 35, 982-1003.

Delaney, J. T. & Huselid, M. A. (1996). The impact of human resource management practices on perceptions of organizational performance. *Academy of Management Journal,* 39(4), 949-969.

Dirksen, V. (2000). The cultural construction of information technology. *Journal of Global Information Management,* 9(1), 5-10.

Dowling, P. J. (1999, October 15). Completing the puzzle: Issues in the development of the field of international human resource management. *Management International Review,* 27-32.

Drucker, P. (1999). *Management Challenges for the 21st Century*. New York: Harper Business.

Ehikhamenor, B.A. (1999). Cognitive information foundation of university students: Index of information and communication technology in Nigeria. *Information Technology for Development,* 8, 139-144.

England, G. W. & Lee, R. (1974). The relationship between managerial values and managerial success in the United States, Japan, India, and Australia. *Journal of Applied Psychology,* 59, 411-19.

Erez, M. (1994). Toward a model of cross-cultural industrial and organizational psychology. In H.C. Triandis, M.D. Dunnette, L.M. Hough (Eds.), *Handbook of Industrial and Organizational Psychology* (Vol. 4, pp. 559-608). Palo Alto, CA: Consulting Psychologists Press.

Evans, P. & Wurster, T.S. (2000). *Blown to Bits*. Boston, MA: Harvard Business School Press.

Feldman, G.M. (1992). Sub-Saharan Africa: Economic reforms are at a critical crossroads in 1992. *Business America,* 113(7), 37-40.

Fernandez, D.R., Carlson, D.S., Stepina, L.P., & Nicholson, J.D. (1997). Hofstede's country classification 25 years later. *The Journal of Social Psychology,* 137(1), 43-52.

The Fourteen Major Trends (1997). *Plunkett's InfoTech Industry Almanac,* 7-15.

Gannon, M. J. (1994). *Understanding Global Cultures: Metaphorical Journey through 17 Countries.* Thousand Oaks, CA: SAGE.

Hampden-Turner, C. & Trompenaars, F. (1993). *The Seven Cultures of Capitalism.* New York: Doubleday.

Hodgetts, R. M., Luthans, F., & Slocum, J. W. (1999). Strategy and HRM initiatives for the '00s environment: Redefining roles and boundaries, linking competencies and resources. *Organizational Dynamics,* 28(2), 7-21.

Hofstede, G. (1980). *Culture's consequences: International differences in work-related values*. London: SAGE.

Hofstede, G. (1991). Management in a multicultural society. *Malaysian Management Review,* 26(1), 3-12.

Hofstede, G. (1993). Cultural constraints in management theories. *Academy of Management Executive*, 7(1), 81-94.

Hofstede, G. (1997). *Cultures and Organizations: Software of the Mind.* New York: McGraw-Hill.

Huck, S.W., Cormier, W.H., & Bounds, W.G. (1974). *Reading Statistics and Research*. New York: Harper & Row.

Husted, B.W. (2000). The impact of national culture on software piracy. *Journal of Business Ethics,* 26, 197-211.

Jason, P. (1997). Banking on computers. *African Business*, 219, 40-42.

Jones, G.K. & Teegen, H.J. (2001). Global R&D activity of U.S. MNCs: Does national culture affect investment decisions? *Multinational Business Review*, (Fall), 1-7.

Kelley, L., Whatley, A., Worthley, R., & Chow, I. (1995). Congruence of national managerial values and organizational practices: A case for the uniqueness of the Japanese. *Advances in International Comparative Management*, 10,185-199.

Kerwin, K., Burrows, P., & Foust, D. (2000). Workers of the World, Log On. *Business Week*, (February)21, 52.

Lim, C.S. & Baron, J. (1997). Protected values in Malaysia, Singapore, and the United States. Retrieved January 4, 2001 from the World Wide Web: http://www.sas.upenn.edu/~jbaron/lim.htm.

Loo, M. (2000). A constrastive analysis of negotiation styles among Malaysian Malays, Chinese and Indians: A practical guide to doing business with Malaysians. 2000 Academy of Marketing Science Multicultural Marketing Conference. Retrieved January 4, 2001 from the World Wide Web: http://marketing.byu.edu/ams/loo.htm.

Mansell, R. & Wehn, U. (1998). *Knowledge Societies: Information Technology for Sustainable Development.* New York: Oxford University Press.

Montealegre, R. (1998). Waves of change in adopting the Internet: Lessons from four Latin American countries. *Information Technology and People,* 11(3): 235-260.

Naughton, K. (1999). CyberSlacking. *Newsweek*, (November)29, 62-65.

Newman, K.L. & Nollen, S.D. (1996). Culture and congruence: The fit between culture and management practices. *Journal of International Business Studies*, 27(4), 753-778.

Odedra, M., Lawrie, M., Bennett, M., & Goodman, S. (1993). Sub-Saharan Africa: A technological desert. *Communications of the ACM*, 35(2), 25-29.

Ohmae, K. (1999). *The Invisible Continent: Four Strategic Imperatives of the New Economy*. New York: HarperCollins Publishers.

Priem, R. L., Love, L. G., & Shaffer, M. (2000). Industrialization and values evolution: The case of Hong Kong and Guangzhou, China. *Asia Pacific Journal of Management*, 17, 473-492.

Ralston, D. A., Gustafson, D. I., Cheung, F. M., & Terpstra, R. H. (1993). Differences in managerial values: A study of U.S., Hong Kong and PRC managers. *Journal of International Business Studies,* 24(2), 249-275.

Ricks, D. A., Toyne, B., & Martinez, Z. (1990). Recent developments in international management research. *Journal of Management,* 16(2), 219-253.

Rigg, M. & Goodman, S. (1992). Mercosur: Reconciling Four Disparate Information Technology Policies. *International Information Systems,* 11(3), 73-86.

Schuler, R.S. & Nogosvky, N. (1998). Understanding compensation practice variations across firms: The impact of national culture. *Journal of International Business Studies*, 29(1), 159-177.

Schwartz, S. H. (1992). Universals in the content and structure of values: Theoretical advances and empirical tests in 20 countries. In M. P. Zarina (Ed.), *Advances in Experimental Social Psychology*, 35, 1-65.

Smith, P.B., Bugan, S., & Trompenaars, F. (1996). National culture and the values of organizational employees. *Journal of Cross-Cultural Psychology,* 27, 231-264.

Steensma, H.K, Marino, L., Weaver, K.M., & Dickson, P.H. (2000). The influence of national culture on the formation of technology alliances by entrepreneurial firms. *Academy of Management Journal,* 43(5), 951-973.

Sunoo, B.P. (1996). The Employee May Be Loafing: Can you tell? Should You Care? *Personnel Journal,* (December), 55-62.

Thompson, J. (1994). Here come the giants. *African Business*, 190, 42.

Triandis, H. (1989). Cross-cultural studies of individualism and collectivism. In J. Berman (Ed.), *Nebraska Symposium on Motivation,* (pp. 41-133). Lincoln, NE: University of Nebraska Press.

Triandis, H. C. (1994). Cross-cultural industrial and organizational psychology. In H. C. Triandis, M. D. Dunnette, & L. M. Hough (Eds.), *Handbook of Industrial and Organizational Psychology,* 4, 103-172.

Trompenaars, F. (1993). *Riding the Waves of Culture*. Chicago, IL: Irwin.

Vitell, S.J., Nwachukwu, S.L., & Barnes, J.H. (1993). The effects of culture on ethical decision-making: An application of Hofstede's typology. *Journal of Business Ethics*, 12, 753-760.

Wright, P. M., Smart, D., & McMahan, G. C. (1995). Matches between human resources and strategy among NCAA basketball teams. *Academy of Management Journal,* 38, 1052-1074.

Wright, R.W. & Ricks, D.A. (1994). Trends in international business research: Twenty-five years. *Journal of International Business Studies*, 25, 687-693.

Youndt, M. A., Snell, S. A., Dean, Jr., J. W., & Lepak, D. P. (1996). Human resource management, manufacturing strategy, and firm performance. *Academy of Management Journal,* 39(4), 836-866.

APPENDIX A

Instrument

These questions in the survey are concerned with your background and work experience.

1.1. Indicate which of the following categories best describes the business or industry your company is in *(please check one)*.

1. □ Manufacturing
2. □ Services
3. □ Wholesale or retail trade
4. □ Finance, insurance, or real estate
5. □ Education
6. □ Self-employed
7. □ Student
8. □ Other __________ *(please specify)*

1.2. What is your best estimate of the number of people who work for your company?

1. □ 2-49
2. □ 50-99
3. □ 100-249
4. □ 250-499
5. □ 500-999
6. □ 1,000-4,999
7. □ 5,000-9,999
8. □ more than 10,000

1.3 How many years have you been employed in this company?

__________ *(to the nearest year)*

1.4. Which of the following categories best describes your current position? *(check one)*

1. □ Top level manager
2. □ Middle level manager
3. □ Lower level manager
4. □ Technical position
5. □ Administrative support
6. □ Other __________ *(please specify)*

1.5. How long have you been in this position?
__________ *(to the nearest year).*

1.6. Please indicate your current salary range?
1. ☐ 2-49
2. ☐ 50-99
3. ☐ 100-249
4. ☐ 250-499
5. ☐ 500-9998.
6. ☐ 1,000-4,999
7. ☐ 5,000-9,999
8. ☐ more than 10,000

1.7. What is the highest level of education you have completed? *(check one)*
1. ☐ High School
2. ☐ Some college
3. ☐ Bachelor's degree
4. ☐ Some graduate or professional study
5. ☐ Graduate or professional degree

1.8. Your age
(to the nearest year) __________

1.9. Gender:
1. ☐ Male
2. ☐ Female

1.10. Does your company have a website?
1. ☐ Yes
2. ☐ No

1.11. Do you have Internet access at work?
1. ☐ Yes
2. ☐ No

Please indicate how likely you would be to access the following types of web pages *while at work*:

1 = Very unlikely
2 = Unlikely
3 = Likely
4 = Most Likely
5 = Very likely

Competitors' website	1 ☐	2 ☐	3 ☐	4 ☐	5 ☐
Government/research websites	1 ☐	2 ☐	3 ☐	4 ☐	5 ☐
General interest websites	1 ☐	2 ☐	3 ☐	4 ☐	5 ☐
Suppliers' website	1 ☐	2 ☐	3 ☐	4 ☐	5 ☐
Customers' website	1 ☐	2 ☐	3 ☐	4 ☐	5 ☐
Arts and Entertainment websites	1 ☐	2 ☐	3 ☐	4 ☐	5 ☐
Travel and Leisure websites	1 ☐	2 ☐	3 ☐	4 ☐	5 ☐
Living/Consumer websites	1 ☐	2 ☐	3 ☐	4 ☐	5 ☐
Business/Financial websites	1 ☐	2 ☐	3 ☐	4 ☐	5 ☐
Sports/News websites	1 ☐	2 ☐	3 ☐	4 ☐	5 ☐

In this section you are asked to give your opinion on the following uses of the *Internet while at work*:

1 = Strongly Disagree
2 = Disagree
3 = Neither Agree nor Disagree
4 = Agree
5 = Strongly Agree

I feel that using the Internet for personal searches is acceptable.
1 ☐ 2 ☐ 3 ☐ 4 ☐ 5 ☐

In my company, it seems that accessing the Internet for personal. searches is tolerated
1 ☐ 2 ☐ 3 ☐ 4 ☐ 5 ☐

My company blocks access to Internet sites which it deemsinappropriate for business use.
1 □ 2 □ 3 □ 4 □ 5 □

Accessing unsecure websites is a potential threat to my company's information system.
1 □ 2 □ 3 □ 4 □ 5 □

My company has incorporated virus protection software.
1 □ 2 □ 3 □ 4 □ 5 □

Internet access increases the risk of importing viruses into my company's system.
1 □ 2 □ 3 □ 4 □ 5 □

My company considers it important to provide its employees with regular reports on their Internet usage.
1 □ 2 □ 3 □ 4 □ 5 □

Please provide us with information regarding the security/monitoring methods adopted your *place of work*:

Does your company block access to certain Internet sites?
1. □ Yes 2. □ No

Do you have additional passwords to access the Internet?
1. □ Yes 2. □ No

My company regularly updates its virus protection software
1. □ Yes 2. □ No

My company has clearly stated Internet usage policy
1. □ Yes 2. □ No

My company strictly enforces its Internet policy
1. □ Yes 2. □ No

Chapter X

Legal Implications of Personal Web Use in the Workplace

Grania Connors
Consultant, Law and Technology, United Kingdom

Michael Aikenhead
University of Durham, United Kingdom

ABSTRACT

The virtues of the Internet as a business tool have been widely extolled: the Internet instantly makes available information that may be difficult or time consuming to obtain by other means. However, use of the Internet in the workplace is fraught with potential problems. This chapter examines the legal implications of personal Web use in the workplace. Specifically, it focuses on the legal issues which can arise for both employers and employees when an employee uses organizational computing facilities for non-work related activities such as surfing the Internet, sending e-mail, chatting online, or instant messaging.

INTRODUCTION

The risk, both legal and otherwise, to employers of liability for employee misconduct is great. Apart from claimed productivity issues and issues with security, personal Internet use in the workplace can result in employers facing claims that: an unsafe or hostile work environment was tolerated, along with associated claims for sexual and racial discrimination, the sharing or selling of confidential information, breaches of trade secrets or corporate intellectual property, claims of defamation, spoliation of evidence, and breach of securities law, to mention but a few.

The precise risks of each of these liabilities requires examination in its own right. The general law will usually apply to many crimes committed via electronic means in the workplace. This chapter does not focus on that general law, but rather on the law relating specifically to an employer's right to monitor employee personal computer use in the workplace, and the rights of employees to limit or avoid such monitoring.

In light of the significant risks to which employers are exposed, it is not surprising that employer monitoring of employee Internet activity is growing at a rapid rate. One study estimates that "[f]ourteen million employees — just over one-third of the online workforce in the United States — have their Internet or e-mail use under continuous surveillance... [while] one-quarter of the global online workforce" is monitored.[1] Such monitoring is usually justified on the basis of property ownership: the employer owns the organization's equipment and networks, therefore the employer has the right to monitor any activity occurring in this environment. This chapter examines the extent to which this property analysis, coupled with vicarious liability issues that employers face, outweighs an individual's right to privacy in the United States.

SOURCES OF LAW

Employers might want to monitor and restrict employees' access to and use of the Web for several reasons. Some of these reasons relate to productivity, other reasons concern limiting an employer's exposure to a range of potential legal liabilities. However, whatever the legality of such behavior, the use of the Web by an employee is a separate matter.

This discussion focuses not on the details for a founding of employer accountability for employees behavior. Rather, this discussion focuses on the related, though distinct issue of whether an employer can monitor and control an employee's use of the Web in the workplace. A corollary of this focuses on the avenues open to an employee who objects to being monitored, or objects to the nature or scope of monitoring to limit that monitoring.

As will be seen, the law governing an employee's personal use of the Web in the workplace is actually an assortment of laws. There is no overarching legal framework that specifies what an employer can and cannot do in every circumstance, and no overarching specification of an employee's legal protections.

The laws affecting an employee's personal use of the Web in the workplace, the right of an employer to monitor and control that use, and the ability of an employee to restrict that monitoring and control is a mixture of federal constitutional law, state constitutional law, employment and labor law, federal electronic privacy statutes, state electronic privacy statutes, federal common law, state common law, and other miscellaneous sources. Despite this range of sources, the law provides little limit on the monitoring and control of an employee's personal Web use in the workplace and consequently provides little power to an employee to limit monitoring they are subjected to or to ease the restrictions that are imposed upon them.

Jurisdiction

As with all legal issues, establishing jurisdiction is centrally important in determining the legal framework that applies. However, although a correct legal basis has to be provided, as concerns employer monitoring of employees, both federal and state laws provide largely the same treatment. The rights of employers vis-à-vis employees are comparatively uniform.[2]

However, jurisdiction is important in an international context. For example, a North American firm conducting business in the United Kingdom will have to comply with different laws and observe different employee rights than the same firm conducting business in its North American offices. Similarly, that firm operating in another European Community state will have to observe different standards than when conducting business in the United Kingdom or when conducting business in North America.

The Constitutional Protection of Privacy

The United States constitution contains no express provision protecting privacy. However, the Fourth Amendment to the Constitution under which:

> *The right of the people to be secure in their persons, houses, papers, and effects, against unreasonable searches and seizures, shall not be violated ...*[3]

does limit the acts to which citizens can be subjected. This protection is limited. This protection of an individuals rights and liberties only applies where there is "state action." This may be present through action by federal, state, or local government, or a branch thereof, however, without such an element no breach of rights under the United States constitution can be maintained. Workers in the public sector may thus have some protections under the constitution; however, even these protections will be limited to what is reasonable and what has been consented to.

The Fourth Amendment has been applied to employer monitoring. In *O'Connor v. Ortega*[4] the Supreme Court held that a state hospital official's search of the office of a physician was unreasonable in the circumstances. The Court stated that the propriety of the search had to be "reasonable" as judged "under all [the] relevant circumstances."[5] Accepting the possibility of an expectation of privacy in the workplace, the court further stated that this expectation "may be reduced by virtue of actual office practices and procedure, or by legitimate regulation."[6] A search will be reasonable when there are legitimate business reasons for the search that outweigh the employee's interests in privacy.[7]

What is reasonable depends on the circumstances in which the search occurs and should be judged by the standard of reasonableness given all these circumstances. In *Ortega* the Court found that Dr. Ortega had a reasonable expectation of privacy in his desk and filing cabinets, and noted that the desk contained personal information including personal correspondence, personal files, personal financial records, and personal gifts unconnected to the hospital.[8]

Despite providing wide latitude for public employers to search employees' workspace, the protections provided in *Ortega* to employees appear further limited when applied to computer monitoring. Providing notice to employees that their computer use may be monitored, their e-mails read, and their computers searched counteracts expectations of privacy. In *United States of America v. Mark L. Simons,*[9] the Court held that a search of a computer used by the defendant did not violate his Fourth Amendment rights because the defendant had not proved a legitimate expectation of privacy in the material searched. The employer had a policy clearly stating that Internet use would be monitored and audited as deemed appropriate, including all file transfers, all websites visited, and all e-mail messages. According to the Court, "[t]his policy placed employees on notice that they could not reasonably expect that their Internet activity would be private." Contrast, however, the Court's finding that absent other evidence, Simons did have a legitimate expectation of privacy in the office in which the computer was situated.[10]

Several other cases have reached the conclusion that a notice informing employees of monitoring or other possibility of audit will preclude a reasonable expectation of privacy. For example in *United States of America v. Eric Neil Angevin,*[11] it was held that a university professor had no reasonable expectation of privacy in an office computer supplied for his use by the university. The university's computer policy explicitly provided that the university could inspect such computers at any time to ensure their appropriate use. In *Albert J. Muick v. Glenayre Electronics,*[12] it was held that an employer could seize and examine the contents of a laptop when the company's computer use policy reserved the right of the employer to inspect the laptop at any time.

Indeed, in *McLaren, Jr. v. Microsoft Corp,*[13] it was found that accessing an employee's e-mails stored in a "personal folder," and which were protected by a password known only to the employee, did not violate the employee's rights to privacy. The court ruled that the employee had no reasonable expectation of privacy in the stored e-mail because, before being stored in these folders, the mail had traveled through the employer's mail system where it would have been accessible to the employer. Moreover, even if a reasonable expectation of privacy could be found, the employer's "interest in preventing inappropriate and unprofessional comments, or even illegal activity, over its e-mail system would outweigh [the employee's] claimed privacy interest in those communications."

Given the highly restricted sphere provided to expectations of privacy when computer use is involved, it is difficult to envisage situations in which monitoring an employees computer use would be a breach of an employee's Fourth Amendment constitutional rights.

For private sector employees, the federal constitution provides even less protection as the element of state action will be missing. Private employers will only be subject to Fourth Amendment claims when acting under or in place of federal agencies.

Numerous state constitutions contain provisions substantively similar to the Fourth Amendment. These state constitutional provisions provide similar wide leeway for public sector employer monitoring of employee Internet use.[14] Interestingly the California constitution extends such constitutional protection into the private sector.[15]

Statutory Protections

As with constitutional protections, federal statutes provide strictly limited protection for employees seeking to restrain employers monitoring their private Web use in the workplace. While numerous bills providing stronger protections for employee privacy have been proposed, no overarching protection exists.[16] The most relevant federal statutes are the Electronic Communications Privacy Act (ECPA) and the Stored Communications Act (SCA).[17] Title 18 of U.S.C. Section 2510 governs any person, including an employer, who intentionally intercepts, endeavors to intercept, or procures any other person to intercept or endeavor to intercept, any wire, oral, or electronic communication.

In the ECPA an:

> *... "electronic communication" means any transfer of signs, signals, writing, images, sounds, data, or intelligence of any nature transmitted in whole or in part by a wire, radio, electromagnetic, photo electronic, or photo optical system that affects interstate or foreign commerce....*

With this broad definition of electronic communication, the Act has been applied on numerous occasions to employer monitoring of employee e-mail and Web browsing.[18] However, in *Steve Jackson Games, Inc. v. United States Secret Service*,[19] the Court ruled that accessing unread e-mail messages was not an "interception" because of lack of contemporaneity with transmission. In *Konop v. Hawaiian Airlines, Inc.*,[20] it was found that accessing a website without authorization does not violate this provision of the ECPA. In *Konop* the Court ruled that to violate the ECPA, the website must be acquired while in transmission and not merely from storage. However, Title 18 of U.S.C Section 2701 governs anyone who:

> *...obtains, alters, or prevents authorized access to a wire or electronic communication while it is in electronic storage.*

While apparently filling gaps highlighted in *Jackson* and *Konop,* the scope of the SCA has been given restrictive interpretation. For example, in *Fraser v. Nationwide Mutual Ins. Co.*,[21] the Court ruled that the SCA did not apply once an e-mail had been received—the SCA only applied to communications in "intermediate storage" or "back-up storage." Compare this with *Konop* where the Court found that the employer had accessed a website without authorization and was thus in breach of the SCA. Both criminal penalties (fines and imprisonment) and a civil action to recover damages are provided in the event of contravention of these provisions.

While prima facie protecting employees' Web use from employer monitoring, the ECPA in fact provides few restrictions for employers. The ECPA contains three exceptions with which employers can claim authorization for their monitoring:

- the provider exception,
- the ordinary course of business exception, and
- the consent exception.

The provider exception renders exempt conduct by an officer, employee, or agent of a provider of electronic communication services which would otherwise breach the Act if the interception occurs during an activity necessary to the rendition of the service or to the protection of the rights or property of the provider.

The extent of this exception is a matter of debate.[22] Commentators:

> *...have predicted that most private employers will be exempt from the ECPA under this exemption if they provide their employees with e-mail service through a company-owned system.*[23]

What it means to be an e-mail system provider and hence within the scope of this exception is unclear; however, an employer would clearly have to provide more to the employee to be classed as an Internet service provider and thus avail him/herself of the provider exception when monitoring private Web use. This is a moot point, however, as the ordinary course of business exception and the consent exception provide broad exceptions for an employer seeking to monitor Web use.

Among other things, the ordinary course of business exception exempts from the operation of the ECPA an employer who uses a device or apparatus to intercept an electronic communication when:

> *...being used by a provider of wire or electronic communication service in the ordinary course of its business.*[24]

An electronic communication service is:

> *...any service which provides to users thereof the ability to send or receive wire or electronic communications.*[25]

Cases raising the ordinary course of business exception typically involve employer monitoring of employee telephone usage. For example, in *Simons v. Southwestern Bell Telephone Co.*,[26] the employee alleged that his conversations, including his private conversations, were being monitored. In ruling for the employer, the Court noted that monitoring was done for quality control purposes and to prevent use of the monitored lines for personal calls, and also that a separate non-monitored phone line was provided for personal calls. The court concluded that the monitoring activities were reasonable and in the ordinary course of business, and therefore covered by the exception.

In contrast, in *Watkins v. L.M. Berry & Co.*,[27] employees were told that personal calls would only be monitored to the extent necessary to determine whether a call was personal or business related. A conversation was monitored in which the employee discussed seeking a job elsewhere. The Court ruled that this was a personal conversation and rejected that "in the ordinary course of business" includes anything of interest to the employer's business. The Court stated that the exclusion does not apply to the interception of personal calls except to the extent necessary to guard against unauthorized use of the telephone or to determine whether a call is personal or business.

When applying the ordinary course of business exception, courts typically apply either the *content approach*, which permits monitoring business-related communications but not personal communications, or the *context approach,* which examines an employer's reason for monitoring to determine whether a legitimate business justification for the monitoring exists. Here factors such as whether the employer provided notice to the employee of the monitoring and whether the level of monitoring was justified are both relevant.

Depending on the circumstances of the monitoring and which test is applied, reading all e-mail, monitoring the content of e-mail, and logging Web use may not be justified under the ordinary course of business exclusion. However, in many cases whether an e-mail or a website is work related will be evident from either the address of the recipient or sender of the e-mail or the Web page. Actually reading a private e-mail may be a breach of the ECPA. However, it would be difficult to argue that actually viewing a Web page address would breach the ECPA since, armed with the Web address, like any member of the public, the employer would be free to view the page.

Whatever are the precise restrictions on the provider exception and the ordinary course of business exception, the consent exception provides employers wide leeway under the ECPA to monitor employee Web use. The ECPA states that a "party to the communication" may give "prior consent" to interception — even when other parties are unaware of the interception.[28]

In *Watkins,* the Court stated that consent:

> *...is not to be cavalierly implied. Title III expresses a strong purpose to protect individual privacy by strictly limiting the occasion on which interception may lawfully take place...[K]nowledge of the capability of monitoring alone cannot be considered implied consent.*[29]

Similarly, in *Deal v. Spears,*[30] the employer monitored an employee's telephone conversations without providing any indication to the conversants that their calls were being monitored. In ruling for the employee, the Court stated:

> *[The employer] did not inform [the employee] that they were monitoring the phone, but only told her they might do so in order to cut down on personal calls.*[31]

Here, knowledge of not only the capability, but also the possibility of monitoring was not sufficient. Contrast *Simons v. Southwestern Bell Tel. Co.,*[32] where the employee was aware of a general monitoring program, used a business-only phone to make a personal call, and where other phones were provided for making personal calls. In *Simons* the Court found that there was consent to monitoring of the calls made.

These and other cases examining the consent exception indicate that for the monitoring of telephone conversations, implied consent will only be inferred where there is clear prior knowledge that communications will be intercepted.

It is therefore anomalous that the consent exception is given less substance when the subject of monitoring is employee e-mail or employee Web use.[33] If an employee is aware that e-mail is being read by the employee and nevertheless continues to send e-mail, then it may be less difficult to infer that they have consented to monitoring of sent e-mail.[34] However, that consent to monitoring will be found even where an express assurance that monitoring will not be conducted is difficult to rationalize with cases examining consent to telephone and other forms of workplace monitoring.

Although not a case examining the consent exception in the ECPA, *Smith v Pillsbury Co.*[35] is widely cited and illustrates the diluted content of the consent exception. In *Smith,* an employee was dismissed after management obtained copies of e-mails sent. Although employees had to log on to the employer's computer system and were warned that e-mail communications were not secure, Smith claimed that his employer had repeatedly informed employees that e-mail would remain confidential and would not be used for the purposes of reprimand.[36] Despite such assurances, the Court found that Smith had no expectation of privacy (and thus by sending messages over the e-mail system consented to them being monitored) — the Court reasoned that because the messages were accessible when they traveled over the employer's network, expectations of privacy could not be maintained.[37] Moreover, the Court determined that even if there was an expectation of privacy, the employer's legitimate need to monitor e-mail might override an employee's interest in privacy. This interpretation of the nature of consent not only appears discordant with cases involving telephone monitoring, but also with privacy law jurisprudence whereby the mere fact that a 'public' facility is involved does not remove the expectation of privacy.[38] Similarly, it is difficult to imagine why an employer's business interests in monitoring e-mail outweigh an employee's expectation of e-mail privacy; such business interests would not outweigh the opening and monitoring of all postal letters written on the employers premises, in an employer-provided office at an employer-provided desk using employer-provided ink. The interpretation in *Smith* of the nature and quality of an employee's consent to e-mail monitoring provides wide scope for operation of the consent exception in the ECPA.

Given the wide interpretation that courts have given to events that will negate an employee's expectation of privacy and hence from which consent can be inferred, it is difficult to imagine situations under which consent will not be found. As seen in *Smith* even an express assurance that e-mail would not be monitored did not negate consent to monitoring. Nor does password protecting computer use. In a case examining consent in the context of tort, *Bourke v. Nissan Motor Corp.,* several employees argued that because they had to use a password to access their computer and their e-mail, and because they were told to keep this password confidential, this raised a reasonable expectation of privacy. The Court disagreed, however, stating that employees knew that e-mail was occasionally read other than by the recipient. Even without knowledge that e-mail is read by a third party, and even password protecting an individual folder marked as 'personal,' an employee may still not have a reasonable expectation of privacy. In *McLaren v. Microsoft Corp.,*[39] another case examining consent to torts, the employee argued that individually password protecting a folder on his computer used to store his e-mail did give rise to an expectation of privacy in this stored e-mail. The Court, however, disagreed, stating that since the e-mails were transmitted over a network where they were accessible to Microsoft before being placed in the password-protected folder meant there was no reasonable expectation of privacy in those messages.

Bohach v. City of Reno[40] is one case examining the scope of the consent exception in the ECPA.[41] In *Bohach* police officers alleged that by retrieving messages sent using their departmentally supplied pagers, their employing police department had violated the ECPA. However, the Court found that the department had warned users that messages would be logged and stored, and would be available for review. As such, the Court ruled that there was no reasonable expectation of privacy and hence that in sending the messages, the officers had consented to them being audited.[42]

On the construction of consent provided in *Bourke*, *McLaren*, and *Bohach*, with notice that e-mail will be monitored, there appear few circumstances in which there can be a reasonable expectation of privacy, and employees appear to have little recourse to prevent monitoring. The use of encryption on messages that are sent and received will possibly raise an expectation of privacy in those messages. Here, there appears a clearer acknowledgment that e-mail is publicly accessible and clear action taken to prevent such access. It is unclear, however, how an expectation of privacy that might be thus raised would be affected by a clause in the employer's acceptable use policy which clearly stated that all decryption keys must be provided for any encryption system that is used.

Similarly, if an employee used a modem built into their personal mobile phone to collect e-mail from a third-party provider or to surf the Web, it is much less clear that such use would be subject to employer monitoring. If an employer-provided computer is used, then the employer could be said to face many of the same business reasons compelling monitoring as existed when the employers own network was used. However, the service provider and ordinary course of business exceptions appear much less compelling. Moreover, the employee has taken steps to remove their behavior from the employer's 'public' network. A prudent employer would require the employee to consent to monitoring of all computer use and files stored on a work computer regardless of how they arrived on that computer. The situation is similarly muddied if an employee uses work resources to access a personal e-mail account (perhaps protected by encryption). The use of a personal e-mail account and encryption would both point towards an expectation of privacy in the e-mail received. However, if this were the case, it would lead to the situation where an employer could audit some of the employees' files, but not others. Again, a prudent employer would expressly require employees to consent to auditing of all their files regardless of their origin.

The rise of tele-working, with its associated blurring of the private and the workplace, raises further legal questions not yet addressed. For example, it is more likely that a tele-worker will use their own computer and their own connection to the Internet (or a connection that is used both for private and work purposes). In such situations, an employer's justification for monitoring is reduced. Moreover, monitoring becomes more fraught as it will be necessary to separate any monitoring of work-related use of the computer and Internet connection from private use by the employee or their family. Regardless of whether or not the employer subsidizes the computer and Internet connection, it seems reasonable to expect more privacy in this situation.

Notably, in the recent case of *Randall David Fischer v. Mt. Olive Lutheran Church et al.*,[43] the Court refused to enter a summary judgment in favor of an employer who had accessed the employee's Web-based e-mail account. Absent any evidence of an acceptable use policy or other indications of consent that the e-mail account would be accessible to the employer, the Court ruled that the employer had a case to answer for a violation of the SCA.

However, discounting cases such as Randall and given employee acceptance of an acceptable use policy, the consent exception is read so widely, and consent given so little substance, that with sufficient notice there has been little that employers have not been able to monitor. As Rosen asks, the question in situations of employee consent is whether an employee can really consent under conditions that are so coercive and where the balance of bargaining power is so uneven.[44]

Common Law

Additional to claims for breaches of constitutional rights, breaches of the ECPA, and breaches of the SCA, monitoring of e-mail and Web usage often raises issues under federal or state common law. Courts in almost all jurisdictions recognize several causes of action addressing intrusion into the personal privacy of individuals: unreasonable intrusion upon the seclusion of another,[45] appropriation of the other's name or likeness,[46] unreasonable publicity given to the other's private life,[47] and publicity that unreasonably places the other in a false light before the public.[48] Of these actions, unreasonable intrusion upon the seclusion of another and unreasonable publicity given to the other's private life are the most relevant for present purposes.

One who intentionally intrudes, physically or otherwise, upon the solitude or seclusion of another or his private affairs or concerns, is subject to liability to the other for invasion of his privacy, if the intrusion would be highly offensive to a reasonable person.[49]

To establish this tort, there must be (a) an intrusion, which is (b) highly offensive to a reasonable person. Accepting that electronic surveillance will constitute "intrusion," courts will look at:

> *...the degree of intrusion, the context, conduct and circumstances surrounding the intrusion, as well as the intruder's motives and objectives, the setting into which he intrudes, and the expectations of privacy of those whose privacy is invaded.*[50]

The expectation of privacy is thus central and an employer's steps to negate any expectation of privacy and an employee's actions to evidence an expectation of privacy are thus central. As with actions under the ECPA and the SCA, acquiescence to an acceptable use policy will often do much to remove such an expectation. Moreover, even with an expectation of privacy established, this tort will not be established if the business reasons for monitoring are so compelling as to outweigh the expectation.

One who gives publicity to a matter concerning the private life of another is subject to liability to the other for invasion of his privacy if the matter publicized is of a kind that (a) would be highly offensive to a reasonable person, and (b) is not of legitimate concern to the public.[51]

The difficulty in establishing this tort is establishing both that the disclosure was highly offensive and that it was of legitimate concern. In terms of monitoring, both involve examination and balancing of the reasons for the monitoring. However, this tort may be relevant to control what use is subsequently made of information that is located during the course of monitoring.

Acceptable Use Policies — Contracts of Employment

The rights of employers and employees in the workplace are often determined by the terms contained in employment contracts. This is so in relation to the issue of personal Web use in the workplace. As between employer and employee, the regulation of personal Web use is largely determined by the contract of employment or whatever industry agreements exist. However, in determining the terms of such contracts, organizational Internet policies are key, since written employment contracts do not usually make specific reference to monitoring of personal Web use. Rather, organizational rules regarding employee Internet use are usually stated in corporate policy manuals. The consequence of this is that rules governing employee Internet use may or may not form part of an employment contract, depending on the manner in which such rules are drafted and the language used therein.

An organization's policy on Internet use forms the nexus between employment and privacy law. The status of enterprise policies regarding Internet use is well established: an employer's policy regarding Web use is usually binding. There have been numerous cases where an organization's employment policies have been held to form part of the terms of an employee's employment contract.[52] Consequently, the terms of a policy may bind the parties to the contract, i.e., both the employer and the employee.[53] This is important in relation to lawsuits that may be brought by disgruntled employees: a breach of organizational policy will often go a long way in proving that an employer is not guilty of alleged misconduct.

This section examines the importance of giving "notice" to employees of an organization's Internet policy, outlines the content that should be included in a policy to reduce employer exposure to various types of liability, and discusses the legal importance of enforcement of Internet use policies.

Enforceability and Notice

For an Internet use policy to apply to an employee or to form part of an employment contract, "notice" must be given of the policy. This means that the terms of the policy are communicated to an employee such that the employee understands that they constitute, in part, the terms on which they are retained to perform their professional duties.

When an employee has notice of an organizational employment policy, they must ensure that their conduct in performing their job conforms to the terms mandated by such policy, just as they are bound by terms expressly included in their employment contract, by statutes prohibiting illegal behavior, or by duties that are generally implied in an employment context, e.g., the duty to perform duties with reasonable care, in a sober state, and without physically harming co-workers.

It is important to note that once an employer has given notice to an employee of the terms of an Internet use policy, the implied consent on the part of the employee that arguably flows from such notice is an important defense to many actions that may be brought against the employer by the employee.[54]

Adequate notice should be given to employees of the terms of an organization's policy.[55] This can be done by circulating the policy periodically via internal memoranda; placing it in employee handbooks, union contracts, or collective bargaining agreements; incorporating the policy into corporate intranets; referring to it at meetings; or placing reminder stickers on workplace computers. It should be noted that privacy rights advocates have argued that, as a bare minimum, a "splash screen" warning should be displayed each time an employee starts their computer, particularly where there is ongoing continuous monitoring of employee Internet use.[56] When considering notice, it is in an employer's interests to over-communicate the contents of its Internet use policy.

Express Versus Implied Consent

Giving employees notice of an Internet use policy is fundamental in establishing consent of employees to review Internet use. Consent, whether express or implied, is a defense to the various constitutional and tortious claims that may be brought against an employer by a disgruntled employee. Employee consent to review will generally be implied from adequate notice of an Internet use policy. Express consent, for example, where an employee signs a written consent or acknowledgment of an Internet use policy, will afford an employer even greater protection from potential lawsuits. To this end, an organization should endeavor to obtain a signed acknowledgment or consent to electronic monitoring for its staff.

Organizational Policy

Preliminary Issues

Given that Internet use policies play such a critical role in the legal regulation of employee Web use in the workplace, it is important to highlight the issues that inform the drafting of such a document, the content that should be included in a policy to effectively limit employer exposure to various kinds of liability, and suggest measures that should be followed when enforcing an Internet use policy.

The starting point in developing and implementing an Internet use policy should be a consideration of the objectives of the policy. An organization should clearly identify what it is attempting to accomplish via the policy. Once the business purpose of such a policy is defined, the content of the policy should be informed by these purposes, and any policy implementation should be tailored to meet them, i.e., a company should avoid excessive or gratuitous monitoring. A company should consider whether different levels of monitoring will be implemented for different categories of staff.

The second step in policy drafting is to identify which categories of personnel to monitor, and which categories of staff will perform the monitoring function. An organization should consider whether monitoring will be carried out by middle managers or network administrators, and whether the actions of these monitors will be subject to review from higher management. As middle managers are often responsible for promotion and discipline of staff, it is prudent to "monitor the monitor" to avoid allegations of discrimination and disparate or selective treatment. In addition, it may be that middle managers are, in turn, subject to monitoring, if they have access to sensitive corporate data.

An organization should also consider its capacity to implement any policy that is enunciated. It is extremely important to ensure that implementation and enforcement of an Internet use policy is uniformly and consistently applied across the whole organization according to the terms of the policy. Failure to enforce sanctions consistently can lead to claims of discrimination and unfair treatment.

Finally, a company should seek legal advice as to whether it is necessary to consult with unions regarding proposed monitoring activities. It may be that collective bargaining agreements or union contracts limit the ability of an employer to monitor workplace Internet activity.

When drafting an Internet use policy, consider the culture of the workplace. Although strict Internet use policies may seem initially attractive vis-à-vis their ability to reduce the risk of employers being held responsible for employee wrongdoing, it is important to consider whether a strict policy will have a detrimental effect on workplace morale, productivity, and cohesion.

Content of Policy

It is important that an organization use clear and specific language in an Internet use policy. The policy should inform employees that:

a) Internet use and e-mail will be monitored;
b) tracking software will be used to monitor Internet use and e-mail;
c) electronic data, including all Internet, e-mail, and instant messaging files, are the property of the employer, not the employee;
d) the employee agrees that electronic data, including all Internet, e-mail, and instant messaging files, are not private and are subject to employer access and review, whether in transit or in storage;
e) the organization's policy applies to all electronic communications, whether sent or received;
f) computer equipment and networks owned by the organization should not be used for personal matters and is provided solely for work-related use;[57]
g) content that is offensive, obscene, profane, indecent, tortious, defamatory, illegal, harassing, or disruptive (including pornographic material or material that is offensive regarding sex, gender, sexual orientation, age, religious beliefs, political beliefs, race, or ethnic origin) should not be created, accessed, or distributed in any way using computer equipment and networks owned by the organization;
h) creating, accessing, or distributing material referred to in paragraph (g) using computer equipment and networks owned by the organization is an act of gross misconduct;
i) confidential information and trade secrets should not be externally distributed in any way using computer equipment and networks owned by the organization without authorization;
j) copyrighted or streaming content material, including software, music, and video programs, should not be downloaded from the Internet using computer equipment and networks owned by the organization without a license to do so and proper organizational authorization;
k) the employee agrees that e-mail accounts used for work purposes may be audited by the employer regardless of whether the e-mail service is provided by the employer or otherwise;
l) the employee agrees any computer used for work purposes may be audited by the employer regardless of the location of such computer and regardless of whether the computer is provided by the employer or otherwise;
m) the employee agrees to decrypt messages when required to do so by the employer for the purposes of business audit;
n) passwords used in respect of company information systems do not imply a right to privacy and do not prohibit an employer from review of any electronic data owned by the company;

o) breach of the organization's Internet use policy should be reported to a designated member of staff as soon as possible; and
p) breach of the organization's policy will lead to disciplinary measures, up to and including termination.

Enforcement and Other Issues

Once an Internet use policy has been drafted, it is important that it is enforced. Firstly, it is important that an organization limits its monitoring of personal communications to the terms of the policy and avoids excessive surveillance that is not performed for a valid business purpose. Employers should implement educational and training programs for management and staff responsible for ensuring compliance with Internet use policies. These programs should stress the importance of vigilance in ensuring that staff members comply with the policy and that policy violations are reported as soon as possible. It is crucial that fair and effective procedures are established and implemented to promptly investigate complaints regarding conduct that breaches the policy, and remedial action should be taken, if necessary. Lax behavior in relation to monitoring or investigation of policy breaches may fuel employee allegations of discrimination or unfair treatment.

An organization should control who has access to information collected as a result of employee monitoring and ensure that such information is disseminated strictly on a "need to know" basis. Careful consideration should be given to what type of filtering software will be used to monitor employee Internet access: the functionality of such software should comply with the scope of monitoring activities disclosed to employees. Data destruction processes should be implemented to ensure that potential Freedom of Information and discovery claims can be effectively managed. Finally, as the law evolves, so should an organization's Internet use policy. A policy should be considered a dynamic document that is regularly reviewed for compliance with employment and privacy laws.

CONSEQUENCES OF BREACH OF LAW

In a recent conference on workplace surveillance, an employer representative identified the risks employers faced by unchecked employee Internet use as including the following:

- misuse of time and capacity in business systems;
- business costs for personal usage;
- harassment of colleagues or outsiders;
- introduction of pornography to the workplace;
- impersonation or malicious altering of documents;
- transmission of viruses;
- inadvertent formation of contracts;
- defamation;
- failure to comply with professional obligations;
- disclosure of commercially sensitive data;
- negligent advice being given;
- loss of patent and other trade secrets;
- breach of copyright and license transgression.[58]

Given the variety and significance of potential employer liability, it is little wonder that employers are attempting to respond to these perceived threats of liability. What's also clear is that employees fired for inappropriate use of the Internet or e-mail will likely consider suing their employers for panoply of perceived wrongs.[59]

This section examines, in relation to employers, employees, and third parties, the consequences of breach of the law as it applies to employee Internet use in the workplace.

As indicated above, the role of an Internet use policy is crucial in reducing an employer's risk of liability for employee Internet use in the workplace. Generally, an organization that has a reasonable use policy that has been publicized to employees and that is conscientiously policed will be in a favorable position to defend a claim brought by an employee.

Cases of extreme misconduct, for example downloading or distributing pornography, will be handled as cases of gross misconduct, and an employer will usually be within their rights to terminate the employment of the responsible individual, subject to adherence to the appropriate warning process.

Invasion of Privacy Suits

As discussed above, Constitutional rights provide a limited basis for breach of privacy lawsuits brought by government employees.[60] However, the majority of cases brought for invasion of privacy are based on the ECPA, the

SCA, and on tort law,[61] where an employer invades the employee's reasonable expectation of privacy.

Courts evaluate the circumstances of a case to decide whether the alleged invasion of privacy is so gross as to render the employee's conduct unacceptable according to the ordinary person standard. Generally, courts have decided in favor of employers, even in cases where employer conduct has breached the contents of their own Internet use policy. *Smith* is an example where, notwithstanding statements that the privacy of personal e-mails would be respected, the court held that there was no invasion of privacy. The company's interest in preventing inappropriate or unprofessional comments or illegal activity outweighed any possible employee privacy interest.

To prove an invasion of privacy claim, the burden is on the plaintiff to prove that a legitimate expectation of privacy exists.[62] The Internet policy can illustrate that no legitimate expectation of privacy exists because the terms of the policy explicitly state that employee Internet and e-mail use in the workplace is not private and is subject to monitoring.

Breach of Federal Wiretap Laws

Breach of the ECPA can result in significant civil and criminal penalties for unlawfully intercepting electronic communications.[63] The scope of the Act covers only "real-time" interceptions of electronic communications — it does not prohibit access of electronic communications in storage.[64]

Although a large proportion of employer monitoring of employee Internet activity will occur in relation to stored files, this will not always be the case. Employers should be aware that real-time monitoring will, in all likelihood, be subject to the provisions of the ECPA. Thus, in the absence of consent to monitoring, software monitoring tools that offer real-time alerting of proscribed activity or dynamically block access to prohibited sites carry a greater risk of falling foul of the ECPA than do software that review activity logs at the end of each day. This should be borne in mind when considering the software solution that will be implemented to enforce an Internet use policy.

As previously noted, monitoring carried out for a legitimate business purpose or where an employee has given express or implied consent will not breach the provisions of the Act. Notifying employees of the contents of an Internet use policy has been held to constitute implied consent to employer monitoring.

Breach of Electronic Communications Privacy Act 1986

The SCA augments the ECPA by prohibiting the access of electronic communications in storage. As indicated above, the service provider exception allows an employer to access all electronic files stored on systems provided by the employer. Even greater employer protection can, however, be found in the ubiquitous "consent" exception: as long as an employer can show employee implied or express consent, they are free to access and review Internet and e-mail files. Implied employee consent to access and review of Internet and e-mail files has been found where an employer informs staff that monitoring will occur.[65]

Harassment, Discrimination, and Hostile Work Environments

Employee electronic communications are commonly used as evidence to substantiate allegations of harassment and discrimination in the workplace. In particular, electronic communications are used to support claims that an employer tolerates a "climate conducive to a hostile environment." Notwithstanding, employers should remember that an effectively enforced Internet policy will result in offensive material being brought to their attention promptly. This ensures that the employer can:

a) reduce the likelihood of being held to be negligent in allowing harassing or discriminatory acts to occur routinely and pervasively;
b) show they acted swiftly and legitimately to remedy any offensive behavior by a worker; and
c) prove that they had a complaints and investigatory procedure by which to address the affected employee's concerns.

These are extremely persuasive in illustrating that an employer did not foster a hostile work environment, took reasonable steps to ensure that such an environment did not evolve, and attempted to minimize detriment to the affected employee.

It should also be noted that courts look for patterns of behavior to conclude that a hostile work environment exists—a single offensive e-mail will usually be insufficient. However, a history of unchecked Internet abuse or recurring circulation of offensive material has been held to suffice in establishing an employee's claim of harassment or discrimination.

At a practical level, it is important that organizations carefully consider the software that they use to monitor employee Internet activity. Software that allows staff to view the Internet activities of their coworkers[66] may prove risky in that it acts as a content publication system within the organization. Unless properly policed, it may be that the organization is routinely responsible for publishing offensive material. This conduct can be used to support a hostile work environment claim.

Public Disclosure: Freedom of Information and Discovery

Both employers and employees should be aware of the fact that Internet abuse carries with it not only the risk of legal liability, but also the possibility that any wrongdoing may be made public. Under Freedom of Information legislation and the procedural rules of court governing discovery,[67] an employer can be forced to hand over electronic data that may be extremely personal in nature or cause severe embarrassment.

Electronic communications are considered to be "documents" according to rules of process. As a result, such documents are discoverable if requested by a party to a proceeding. "By tracking and storing a detailed audit trail of employee activities, organizations may be inadvertently stockpiling large amounts of potential evidence that could be used against them in future litigation."[68] A data retention policy can be useful in limiting information that is available to be produced pursuant to discovery, and can also ensure that the amount of data that must be reviewed in order to comply with any request for discovery is manageable. It should also be noted that information that is "deleted" might often be recoverable.[69] It is therefore prudent to engage in systematic "permanent retirement" of electronic data.[70]

Under the federal Freedom of Information Act and similar state legislation, any third party[71] can obtain the disclosure of public sector employees' personal Internet and e-mail files. This is because the Internet and e-mail files of government employees fall within the definition of "public records." This situation can be contrasted with that of employees in the private sector, who are not subject to Freedom of Information legislation.

It is important to note that some states have an exception to disclosure if records contain personal information. However, in states where there is a personal information exception, disclosure may still be required if there is a valid public interest in the contents of employee records. Furthermore, many state open-records statutes do not contain a personal information exception. If an exception to disclosure cannot be proved by a government employee, they will be hard pressed to avoid public disclosure of Internet or e-mail files, notwithstanding that such files may be extremely personal in nature and highly embarrassing.

Caution: You Can Be Too Careful

Employers should note that any perceived breach of corporate policy regarding Internet use should be approached in a reasonable and balanced manner. There have been a number of cases, for example sending sexually suggestive e-mail to a consenting coworker or "inadvertently" accessing pornographic sites, where employees have been awarded significant damages or have obtained sizeable settlements for discrimination or unfair dismissal. When managing the sensitive area of Internet misuse, caution is advocated.

INTERNATIONAL COMPARISON

Multinational employers or employers that transact with organizations outside the United States should be aware of privacy and data protection laws that operate in foreign jurisdictions.

European employers are bound by strict data protection laws[72] that govern the collection, storage, and transfer of workers' personal information. "These protections place employees on a more equal footing while allowing employers to monitor for legitimate reasons."[73]

Of note, in the United Kingdom, the Human Rights Act 1998 guarantees employee privacy. However, due to employer lobbying, this Act was modified by the enactment of the Regulation of Investigatory Powers Act 2000, which allows employers to monitor employee electronic communications for the purpose of establishing the existence of facts or to ascertain compliance with regulatory or self-regulatory practices and procedures. Under this Act, employees must be informed that monitoring is taking place. In addition, the Data Protection Act 1998 defines standards that apply in handling collected information and how such information should be stored.

In 2000, Australia introduced the Privacy Act 2000, a "light touch" regulatory framework that covers both public and private sectors. However, this Act expressly excludes employee records from its scope.

In 1996, the International Labor Organization (ILO) published a three-volume work, entitled the "Conditions of Work Digest." The Digest contains a code of practice on the protection of workers' personal data that states that workers' data should be collected and used according to Fair Information Practices.[74] Similarly, the OECD has published guidelines regarding employee personal data based on Fair Information Practices.[75]

Finally, many international jurisdictions differ from the United States in that they have created an office of the Privacy Commissioner. For example, several European countries, Canadian, Australia,[76] New Zealand,[77] Japan, and Hong Kong[78] have established this office or an equivalent.

CONCLUSION

Given the history of workplace privacy jurisprudence in the United States, the wide freedom which employers are given to monitor employees and the converse sparsity of avenues of recourse to employees subjected to monitoring is surprising. Neither the federal Constitution, nor state constitutions provide substantial protection to employees from employer monitoring of their Internet activities. Federal and state statutes regulating employer monitoring of employee Internet usage and relevant tort law are similarly permissive towards monitoring. Courts regard monitoring as acceptable when consented to by an employee, and Courts have read the requirement for consent so insubstantially as to render the consent requirement near negligible. Nevertheless, prudent employers will reduce the risk of potential problems with their monitoring activities through adoption of an acceptable Internet use policy and the promulgation of such policy through their organization. Given that monitoring can extend beyond the individual employee and can cover other businesses which an employer retains, the need for such policy heightened.[79]

The legal treatment of monitoring of employee Internet usage in the workplace is all the more surprising given the protection afforded to employee offices and desks, and to the respect for privacy evidenced in restrictions on the monitoring of employee telephone calls. At a time when e-mail is becoming a ubiquitous form of communication, it is questionable whether this inconsistency in treatment is justifiable. However, in a political and legal environment in the United States which has given birth to the USA PATRIOT Act80 with the broad surveillance powers that is establishes and the atmosphere of surveillance it evidences, it is questionable whether employees will see improvement in their situation in the near future.

ENDNOTES

1 Schulman, A. (2001). *The Extent of Systematic Monitoring of Employee E-Mail and Internet Use.* Privacy Foundation: Workplace Surveillance Project, July 9. http://www.privacyfoundation.org/workplace/technology/extent.asp. A previous study conducted by the American Management Association, *Workplace Monitoring and Surveillance*, 2001, found that 77.7% of major United Stated firms monitor employee Internet use.

2 California and Wisconsin are two notable exceptions.

3 Constitution of the United States of America, Fourth Amendment. This applies to states through the Fourteenth Amendment.

4 480 U.S. 709 (1987).

5 Ibid. 725-26.

6 Ibid. 717.

7 Ibid. 719-20.

8 Ibid. 718.

9 206 F. 3d 392 (4th Cir., 2000).

10 The search of the office did not breach the defendant's Fourth Amendment rights, however, because an employer is allowed to conduct warrantless searches as part of investigations of work misconduct, provided that such search is reasonable in scope.

11 No. 01-6097 (10th Cir., February 22, 2002). 10th Circuit.

12 280 F.3d 741, No. 98 C 3187 (7th Cir., 2002).

13 Case No. 05-97-00824, 1999 Tex. App. Lexis 4103 (Tex. Crt. Of App., May 28, 1999), from: http://courtstuff.com/cgi-bin/as_Web.exe?c05_99.ask+D+10706510. Although not a case examining Fourth Amendments rights but rather, involving common law torts in the invasion of privacy issues surrounding this expectation of privacy are extremely similar.

14 Note that several state constitutions explicitly protect privacy and offer greater protection than does the United States Constitution: Ciocchetti, C. (2001). *Monitoring Employee E-Mail: Efficient Workplaces v. Employment Privacy.* Duke L. & Tech. Rev. 0026 at ¶10.

15 Ibid.

16 The Notice of Electronic Monitoring Act (the NEMA) was proposed legislation dealing with how often employers must inform their employees about electronic monitoring introduced by Senator Charles Schumer (D-NY) and Rep. Bob Barr (R-GA) during the 106th Congress, (H.R. 4098/S.2898). Under NEMA, notice would have needed to describe the form

of computer use being monitored, the method of monitoring, the information obtained by the monitoring, and how the information is to be stored, used, and disclosed. Similar bills have been proposed in many state legislatures.

17 Most states have legislation which, with rare exception, mirrors the operation of the ECPA and the SCA.

18 Refer to the cases cited infra.

19 36 F.3d 457 at 461.

20 302 F.3d 868 (9th Cir. 2002).

21 135 F. Supp. 2d 623. C.f. *Randall v Mt Olive Lutheran Church* where the Court stated it "is unnecessary to determine whether Fraser held correctly that the act could be violated only by accessing e-mail that has not yet been downloaded to the recipient's hard drive."

22 Ciocchetti, above n. 14.

23 Ibid. ¶15.

24 18 U.S.C. Section 2510(5).

25 18 U.S.C. Section 2510(15).

26 611 F.2d 342 (10th Cir. 1979).

27 704 F.2d 577 (11th Cir. 1983).

28 18 U.S.C Section 2511(2)(d).

29 704 F.2d 577 at 581.

30 980 F.2d 1153 (8th Cir. 1992).

31 Ibid at 1155-6.

32 452 F. Supp. 392 (W.D. Okla. 1978), aff'd., 611 F 2d 342 (10th Cir. 1979).

33 Dixon, R. Windows Nine-to-Five: *Smyth v. Pillsbury* and the Scope of an Employee's Right of Privacy in Employer Communications, 2 Va. J.L. & Tech. 4 (Fall 1997).

34 *Bourke v. Nissan Motor Corp.* No. B068705 (Cal. Court of App., 2nd Dist., 1993).

35 914 F. Supp. 97.

36 Ibid.

37 Ibid. 101.

38 Dixon, above n. 33.

39 Case No. 05-97-00824, 1999 Tex. App. Lexis 4103 (Tex. Crt. Of App., May 28, 1999).

40 932 F.Supp. 1232 (D.Nev. 1996).

41 *Bohach* involved the consent exception to the accessing of store electronic communications; however, there is no reason to regard the legal nature of the consent exception in this area as different from the consent exemption to accessing electronic communications while in transmission.

42 Moreover, the Court ruled that in supplying the equipment, the police department also fell within the system provider exception.

43 01-C-0158-C (W.D. Wis., March 28, 2002), available at http://pacer.wiwd.uscourts.gov/bcgi-bin/opinions/district_opinions/C/01/01-C-158-C-03-28-02.pdf.

44 Rosen, J. (2000). *The Unwanted Gaze.* City: Random House.

45 Restatement (Second) of Torts 652B (1977).

46 Ibid. 652C.

47 Ibid. 652D.

48 Ibid. 652E.

49 Ibid. 652B.

50 *Miller v. National Broadcasting Co.*, 187 Cal. App. 3d 1463, 1483-84 (Cal. Ct. App. 1986).

51 Restatement (Second) of Torts 652D.

52 *Panto v Moore Business Forms, Inc* 130 NH 730 (1988); *Butler v Walker power, Inc* 137 NH 432 (1993).

53 *Smyth v Pillsbury* C.A No. 95-5712, U.S. District Court for the Eastern District of Pennsylvania, Jan 18, 1996, Decided Jan 23, 1996.

54 This issue is discussed in full below, in the section entitled "Consequences of Breach of Law."

55 Note, privacy rights activists have questioned whether "the practice of keeping employees uninformed about the details of monitoring [is]...tantamount to entrapment." See Schulman, A. (2001). *The Extent of Systematic Monitoring of Employee E-Mail and Internet Use.* Privacy Foundation: Workplace Surveillance Project, July 9. http://www.privacyfoundation.org/workplace/technology/extent.asp.

56 Ibid.

57 This paragraph follows the "strict" model of Internet use policies in that it disallows all personal use of workplace computing facilities. However, as indicated above, it may be that organizations prefer a more lenient policy of limited personal use of workplace computers, provided that employees adhere to policy prohibitions regarding indecent or offensive material. In addition, it is important to note that the National Labor Relations Board has issued an advice memorandum stating that an Internet use policy which

prohibits all non-business use of workplace e-mail may breach the *National Labor Relations Act* (NLRA) if it restricts or interferes with union activity. If an organization's workforce is unionized, its Internet use policy should be informed by this opinion. See *Pratt v Whitney* 26 AMR 36322, 12-CA-18446 (Feb 23, 1998). See also NLRA, 29 USCA sections 151-169.

58 Low, S. (British Chambers of Commerce). (2001). In *Monitoring in the Workplace Conference: Report.* Manchester, June 28.

59 Steinberg, M. & Azar, D. (2001). A watched worker. *Computerworld,* (May 14). http://www.computerworld.com/news/2001/story/0,11280,60407,00.html.

60 See, for example, *Griswold v Connecticut* 381 U.S. 479, 85 S. Ct 1678 (1965) where an implied right of privacy was found, and *O'Connor v Ortega*, where the Supreme Court held that a reasonable expectation of privacy existed in a government workplace.

61 The most common tort actions brought in this context are "unreasonable intrusion upon the seclusion of another" and "unreasonable publicity given to another's private life."

62 *United States v. Rusher* 966 F.2d 868 at 874 (4th Cir. 1992).

63 For illegal interception of electronic communications, a successful claimant may obtain: (a) injunctive relief, (b) damages of $10,000 or $100 for each day of illegal interception, (c) punitive damages, and (d) legal fees and costs. For illegal access of stored electronic communications, statutory damages are capped at $1,000, while punitive damages cannot be recovered.

64 In *United States v Mark L. Simmons* 206 F. 3d 392 (4th Cir., February 28, 2000), the Court held that e-mail allegedly obtained illegally was not technically intercepted, but rather was obtained from storage. As a result, federal wiretapping laws did not apply to render the e-mail access illegal. In *Steve Jackson Games v U.S. Secret Service* 36 F.3d 457 (5th Cir. 1994), it was held that the definition of electronic communications excludes such communications while in electronic storage, and only applies to communications obtained while in transit.

65 Electronic Privacy Information Center, Workplace Privacy, 21 November 2002, http://www.epic.org/privacy/workplace.

66 Such as Fatline's product Fasttracker and AltaVista's product AV Enterprise Search.

67 Discovery is the formal legal process by which parties to a legal proceeding gather evidence from each other. There are strict rules for the

preservation of evidence and delivery of evidence once proceedings have been initiated.

68 *One-Third of U.S. Online Workforce Under Internet/E-mail Surveillance.* http://www.privacyfoundation.org/privacywatch/report.asp, July 9, 2001.

69 In a case involving unfair dismissal, an e-mail specialist located a "deleted" e-mail message from the company's president directing a human resources officer to "get rid of the tight-assed bitch." The case was settled for a considerable sum. See McNeil, H. & Kort, R. (1995). Discovery of e-mail: Electronic mail and other computer information should not be overlooked. *Oregon State Bar Bulletin*, (December).

70 It is important that destruction of data is performed in a *routine* and *systematic* manner to avoid claims that data was destroyed solely for the purpose of frustrating discovery attempts.

71 In cases where the press has been successful in obtaining records pursuant to Freedom of Information legislation, such records have been publicized through state and national media channels.

72 See *Directive 95/46/EC of the European Parliament and of the Council of 24 October 1995 on the protection of individuals with regard to the processing of personal data and on the free movement of such data; Directive 97/66/EC of the European Parliament and of the Council of 15 December 1997 concerning the processing of personal data and the protection of privacy in the telecommunications sector; Regulation (EC) No 45/2001 of the European Parliament and of the Council of 18 December 2000 on the protection of individuals with regard to the processing of personal data by the Community institutions and bodies and on the free movement of such data; Directive 2002/58/EC of the European Parliament and of the Council of 12 July 2002 concerning the processing of personal data and the protection of privacy in the electronic communications sector; Treaty on the European Union, Title I - Common Provisions - Article F* and *European Convention for the Protection of Human Rights and Fundamental Freedoms, Article 8.*

73 Electronic Privacy Information Center, Workplace Privacy, November 21, 2002. http://www.epic.org/privacy/workplace.

74 Note: this code is not enforceable nationally or internationally. Rather, it is intended as a reference when legislation, union agreements, and work policies are being drafted.

[75] Again, these guidelines are not enforceable nationally or internationally. Rather, they are intended as a reference when legislation, union agreements, and work policies are being drafted.

[76] In March 2000, the Australian Federal Privacy Commissioner published "Guidelines on Workplace E-Mail, Web Browsing and Privacy."

[77] The New Zealand Privacy Commissioner has published privacy information at http://www.knowledge-basket.co.nz/privacy/semployf.html.

[78] In March 2002, the Hong Kong Privacy Commissioner published a "Draft Code of Practice on Monitoring and Personal Data Privacy at Work."

[79] Note *Andersen Consulting L.L.P. v UOP* 991 F. Supp. 1041 (N.D. Ill. 1998), a situation where e-mails sent by Andersen over a client network were audited and subsequently published in a national newspaper to the arguable detriment of Andersen. Without authorization for such auditing and subsequent publication, the consequences to the employer of these actions could have been great.

[80] Uniting and Strengthening America by Providing Appropriate Tools Required to Intercept and Obstruct Terrorism (USA PATRIOT ACT) Act of 2001.

Section III

Toward the Well-Being of the Employee

Chapter XI

A Psychoanalytic Perspective of Internet Abuse

Feng-Yang Kuo
National Sun Yat-Sen University, Taiwan

ABSTRACT

In this chapter I discuss Internet abuse from a psychoanalytic perspective. Internet abuse refers to the misuse of the Internet that leads to deterioration of both public and individual welfares. While past research has treated most computer abuse as the result of conscious decisions, the school of psychoanalysis provides insight into how the unconscious mind may influence one's abusive conduct. Therefore, I argue that effective resolution of Internet abuse requires the knowledge of the unconscious mind. Although modern knowledge of this domain is still limited, I believe that this orientation is beneficiary to the construction of social systems embedding the Internet and their application to our work.

INTRODUCTION

Today, we live in a wired society where information technologies have permeated every part of our lives. While we have cherished this achievement, we are also becoming increasingly vulnerable to various forms of computer abuse that infringe upon our basic rights of freedom of speech, privacy, properties, etc. To the professional IT managers in Taiwan, this abuse seems especially troublesome because of the Internet's huge popularity and its negative image portrayed by Taiwanese' public media. Indeed, the sorts of abuse that are seen in the newspapers almost daily are no longer matters like flaming and defamation, which we may call "Internet abuse in the small." Instead, the abuse is much broader in scope socially — gang fighting, broken families, wholesale piracy, and even murders, which we shall call "Internet abuse in the large." Should the company be held liable to those abuses, be they large or small, when employees utilizing the company's computing resources to commit them? To the IT professional managers, curbing Internet abuse becomes a new challenge because they are no longer dealing with problems that they can address with isolated intra-company policies. Rather, Internet abuse in the workplace is intricately linked to the world outside the company. The sources of the abuse are societal and the challenge to understand them seems insurmountable.

Taiwan is rather unique in the adoption of the Internet. Its number of Internet users has grown from 400,000 in 1996 to an estimated six million by the end of 2000, according to statistics released by Taiwan's semi-official Institute of Information Industry (III, 2001). Over half of this Internet population are 30 years or younger, while another quarter belongs to the 30-something group. Almost two-thirds are college educated or equivalent, and over half access the Internet daily. One would think that such a population profile points to a healthy picture of Internet usage. Yet, according to YAM (http://www.yam.com), the civil watchdog of Taiwan's Internet, the most popular websites in 2000 are consistently services in which illegal transactions of sex, computer software, movies, and drugs are likely to be conducted. Furthermore, in the year 2000, more than 90% of news pertaining to the Internet reported in the public media was negative, such as wholesale software piracy, sex trades, broken families, and gang fighting.

The Internet has been portrayed as the core engine empowering us to a state of the ultimate democracy and the friction-free (transaction cost-free) market. But in Taiwan, while none of these virtues are in sight, the society is seemingly already paying a price for this technology. Is this only a temporary but necessary step before transition into a better future? Or is the future already here? Or is information technology, however powerful it might be, only a slave of the culture in which it is implemented?

These are difficult questions to answer. They are difficult because the Internet is itself an evolving technology. They are difficult also because we don't seem to be equipped with adequate knowledge to study it. Past research of the Internet has been based on theories of rationalistic tradition and has focused mainly on the possible positive contributions. Yet, as revealed above, many Internet abuses, especially those in the large, are beyond the power of rationalistic theories to explain. Thus, this chapter attempts to evaluate Internet abuse from a psychoanalytical perspective. In the following, two important theories of psychoanalysis, Freud's structural model and Sullivan's interpersonal integration, are discussed. A case study of a class of professional IT managers is then presented, and the implications of this case study are discussed, followed by the conclusion.

THE THEORIES OF FREUD AND SULLIVAN

How could the concepts of psychoanalysis, already a century-old and somewhat out of fashion, be related to one of the most advanced achievements of modern mankind, information technology, and its application to our human society? The possible linkage is the human mind, notably the "abusive" conducts resulted from the unconscious, dysfunctional mind that contains an unpleasant past memory. Freud was best known for his work on psychologically disturbed patients, who were physically fit and yet exhibited hysterical symptoms. (For instance, the patient's hand or legs were fine but could not write or walk.) In studying these patients, Freud came to many startling conclusions concerning neurosis. First, the problems were not in the flesh but in the mind. Freud believed that the mind had in itself an unconscious component that constituted an indispensable part of the mental life. Next, Freud reasoned, the mind was an apparatus for discharging stimuli that impinged on it. Central among the stimuli were the instinctual drives of sexuality and aggression. Furthermore, the charged experiences in early life, particularly if they were repressed, might result in serious psychic pathology in later life. Finally, Freud developed the structural model of the mind (see Figure 1) comprising three agencies: id, ego, and superego. The id contained the raw, unstructured,

Figure 1. Freud's Structural Model

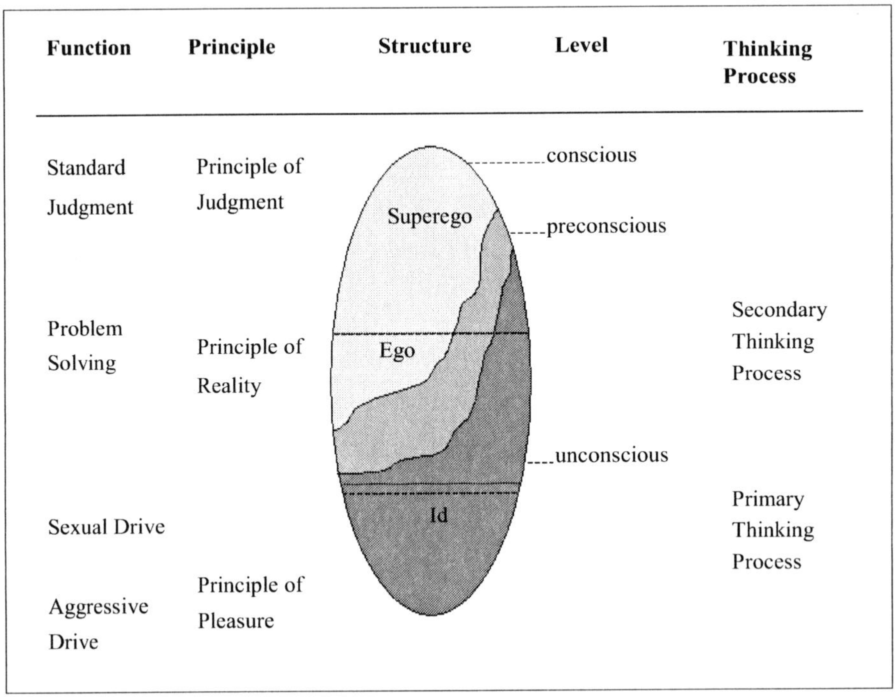

impulsive energies. The ego regulated the mind so that the primitive impulses of the id could be controlled. In the superego was a set of moral values and self-critical attitudes. Furthermore, influenced by Darwinian metaphors of his day, Freud hypothesized that humankind was still evolving and torn by a fundamental rift between bestial motives and civilized conducts. Thus, people were driven to satisfy the id, which led to pleasure, and yet, in order to be acceptable socially, they must also conceal from themselves these purely hedonic motives. The human mind was therefore full of conflicts that are unknown to the mind itself. With the aid of social guidance that is implanted into the superego, the ego can then repress and regulate the primitive impulses of the id. Abusive conducts might occur when the delicate working of one's id, ego, and superego is out of order.

In terms of Internet abuse, misconducts take place most often at the time when people are dis-inhibited or de-individuated. It appears that this is also the time that people's superego is resting. Accordingly, could the abusive conducts be the work of the id where one's instinctual impulses reside and are ready to come out once the superego is absent? The answers would have great implications to how Internet abuses can be approached and resolved.

Freud proposed his theories almost a century ago. Since then, the school of psychoanalysis has gone through a lot of changes. Many neo-Freudians have revised the model of mind. One revision by Harry Sullivan is fundamental and is particularly relevant to societal abuses of the Internet in Taiwan. Unlike Freud who sees the self as isolated from the world and the mind as the captive of the primitive impulses, Sullivan sees people as fundamentally social and the mind as generated in interactions among individuals. This is a radical departure from Freud's original formulation. People have basic needs of integration with others. Satisfying these needs produces pleasure, while the lacking leads to anxiety. People therefore are driven away from anxiety-increasing activities toward pleasure-increasing ones. Finally, the relative enduring patterns of recurrent situations in which one finds pleasure or suffers anxiety will shape the development of his or her self system. Personality defined this way is no longer some innate predisposition, but the product of the history of one's interactions with others. Thus, abusive conducts can occur when the environment is deemed as anxiety laden. Accordingly, curbing abuses will require much more than designing reward/penalty incentives directed at individuals.

As discussed earlier, in Taiwan almost half of the top 10 websites are consistently chat sites, and over half of the user population are rather young. Why are these young folks so attracted to these sites? Can their interpersonal life in the physical world be so dissatisfying that the Internet chat sites become a safe heaven? This possibility that in Taiwan this "abuse in the large" could be attributed to cultural practices of the physical world is somewhat speculative, but not entirely unfounded. In this Confucian society, academic excellence often means a hardworking child studying in isolation. The interpersonal life outside family is discouraged, if not penalized. Can the lack of social integration arouse anxiety and lead to the sort of Internet abuse in Taiwan? The professional IT managers are now at a loss. The problem is too complex to be analyzed.

In a course entitled "Information and Society," a group of students set out to discover answers to these questions. They were not young college students, but professional managers holding mid- to high-level positions in charge of implementing information technologies in their respective companies. Rather than trying to turn them into knowledgeable psychologists and sociologists, the instructor handed out an assignment: get in those popular sites and practice what the youngsters do. The professionals were asked to assume a much younger Internet identity. Furthermore, they must disguise themselves as the opposite sex, i.e., men must disguise as women so that they (who are old men) could learn about what young men do in those Internet chat services, and vice versa. They were asked to record and reflect on their experience upon which their classmates would also conduct a collaborative interpretation. This approach was an implementation of the ethnography for learning from history (Kleiner & Roth, 1997). As it turned out, the assignment was a much more difficult mission than originally expected.

THE MISSION IMPOSSIBLE

The group of IT professionals, averaging about 40 years old, encountered severe difficulty from the very beginning. The two oldest, already in their fifties, decided to be 41-year-old women. To them, the age of 41 is "already young enough," despite they have learned that most Taiwanese Internet users were in their teens and twenties. Later, these two, along with many others who were male, feared and refused to assume a female identity. The female professionals, on the other hand, had similar trouble assuming the male identity. There were more challenges afterwards: they could not speak the lingo of the young and therefore found few people to chat with. They ventured into different websites, some of which were known for their sexual orientation, without much success. Finally after many attempts (and hours), some were able to enter dialogs successfully. And what they discovered was shocking to them. Confirming to the stereotypical image portrayed in the public, there were indeed some very abusive behaviors, which, in Freud's terms, appear to be the work of the impulsively primitive id. For example, one encountered a situation in which he (who assumed a female Internet identity) was asked to have cyber-intercourse. But many other times, the youngsters at the other end of the chat were only seeking integrating relations, as Sullivan would have envisioned. The worst abuse by these youngsters themselves might be that they spent way too much time in Internet chat.

However, the fact that Internet becomes a safe heaven for the young has important implications to the "abuse in the large." For one, their social well-being may decrease (Mitchell & Black, 1995). This reduction in social well-being may in turn affect the physical life of these people, i.e., losing interest in school or in work. Also, many like to bring their Internet discoveries (e.g., pornographic, hatred materials, etc.) to share with their colleagues and friends inside the company or school. This, however, weakens the defense of the company/school, and the potential for moral hazards increases. Finally, their lack of social experience makes them susceptible to criminal acts. The possibility that criminals are lurking around the Internet is nothing new. False advertising is virtually impossible to prevent, and criminals can certainly disguise themselves easily.

Through the assignment, the IT professional managers start to understand that Internet abuse in the workplace is inseparable from the entire ecology of the Internet itself. However, the real surprising discovery for the IT professionals is not about the Internet abuse, but about their own unconscious mind. For example, consider the two oldest male managers who chose to assume the identities of 41-year-old women. When asked the reasons behind their choice, they explained that only lonely mid-aged women would become a frequent visitor of those chat sites, and they had to be "bad" (i.e., acting seductively) for any man to talk to them on the Internet. Unconsciously, their actions revealed several implicit beliefs that are held commonly by many Taiwanese men of their age. First, the Internet is bad, full of sexual and pornographic materials. Next, "normal" women (i.e., "good women") have no use for the Internet. Third, divorced women in their thirties and forties are lonely and vulnerable and likely to become Internet users. Finally, these women must behave in a seductive way for any man to be interested in talking to them.

These negative stereotype beliefs about both the Internet and women are not manufactured by any individual, but are embedded in cultural practices that have existed for a long time. These beliefs, like Brown and Duguid (2000) suggested, are typically undetectable unless there is a breakdown in carrying out actions intended by these beliefs. And indeed, the two professionals would not have admitted to their biases unless their attempts to socialize themselves in the Internet failed. (They failed in the sense that they were not successful in entering a dialog.) The words of the prominent organizational sociologist, Karl Weick, *"How can I know what I think until I see what I say,"* seem to echo (Weick, 1979). They now see what they have done and realized that, in a Freudian sense, they were unconscious of these beliefs that are deeply buried in their mind. It is those beliefs that drive their actions, despite taking courses that teach all the positive applications of the Internet. Furthermore, both have the first-hand knowledge of women who use the Internet: their daughters, in their twenties, have used it often. They should have known better (about both the Internet and female Internet users), but in reality they didn't.

Finally, according to Freud and many later Neo-Freudians, people's behavior is fundamentally couched in the pleasure principle. Thus, satisfying the id's primitive impulses produces pleasure; sometimes people may even seek pleasure to the extent that they become despondent in other aspects of life. Those professional managers' experience seems to suggest that the Internet may indeed be a vehicle for satisfying the need for interpersonal integration, as implied by Sullivan's interpersonal field theory. But there is a subtle, though important difference: this satisfaction is more from the person's own imagination than the real-world socialization. One's imagination, of course, is highly error prone. Thus, befriending on the Internet is like opening the floodgate for hazards, since there is infinite possibility of fidelity that the Internet could provide. Could this lead to abuse or even addiction?

Even more questions linger. Where do the stereotype beliefs come from? Is it true that people are more or less unconscious of these deep beliefs behind their abusive acts when they are de-inhibited or de-individuated? In those isolated situations, if people may act abusively without knowing that their unconscious mind is the culprit, what can the management do to successfully prevent abusive conducts? If people can act against their knowledge (as these two oldest professionals have done), what sort of education can be effective to change the unconscious mind? For those IT professionals, there seemed to be an unlimited number of questions emerging after this assignment.

SCHOOLING THE UNCONSCIOUS MIND: FROM PSYCHOANALYSIS TO COGNITION

If the origin of "abuse in the large" can be traced to the unconscious mind and its surrounding culture, how can it be schooled? The answer to this question is no doubt of great interest to both ethical theoreticians and practitioners. According to the framework laid out by Freud and neo-Freudians, the emotional life of the young child is critical and the remedy resides in the opening up of the unconscious's unpleasant memory. The IT professionals would have no use for this advice since they have no proper training to conduct psychiatric treatment, which would also be too expensive for the company to afford. Fortunately, modern scholars of cognitive psychology have worked on this issue so that we now have some clues on the approaches to schooling the unconscious mind.

Howard Gardner is one of these scholars who have important insight into this matter. Trained in both the Freudian school and the modern cognitive tradition, the prominent Harvard professor of educational psychology has invented the term "unschooled mind," referring to the set of cognitive capacities that one acquires before the age of five. Gardner's research discovers that, before the time of schooling, a person already holds firmly many beliefs about the nature of the world as well as conceptions about people, family, and society. These "unschooled" beliefs and conceptions would become very difficult to be updated by formal schooling. "...In nearly every student there is a five-year-old "unschooled" mind struggling to get out and express itself" (Gardner, 1991, p. 6). Except in fields in which a person becomes an expert, the educated mind, which is filled with various sorts of declarative knowledge that one learns in the school or from books, is losing out to the unschooled one. This view that the human mind may be unschooled has also been observed in business practices, in which "young or old, female or male, minority or majority, wealthy or poor, well-educated or poorly-educated" are all engaged in "Model-I theories in use" that are inconsistent with their declarative beliefs (Argyris, 1990, p. 13). Simply put, in the workplace, persons often say one thing (beliefs that they learn formally) while doing another (in accord with their unschooled theories), and they are not aware of this inconsistency.

This unschooled-ness of our mind has challenged researchers investigating human practical use of information technology in the most fundamental way. For ethics researchers, no quick fix is in sight. But the works of Susan Harrington (1996) and Banerjee et al. (1998) reveal a clue: both demonstrate the importance of the organizational context in which the ethical conduct is taken. For instance, people tend to act ethically in a caring environment. Note that this context interacts with the self system in a reciprocal way. Bandura (1991), in a landmark paper titled, "Social Cognitive Theory of Moral Thought and Action," has elaborated on this reciprocality. Briefly, transgressive conduct is regulated by both social sanction and internalized self-sanction that operate concurrently and anticipatorily. In control arising from social sanctions, people refrain from transgressing because they anticipate that such conduct will bring them social censure and other adverse consequences. In self-reactive control, they behave pro-socially out of self-satisfaction and self-respect, and they refrain from transgressing because such conduct will give rise to self-reproof (Bandura, 1991). The stronger the perceived self-regulatory efficacy, the more perseverant people are in their self-controlling efforts and the greater is their success in resisting social pressures to behave in ways that violate their standards. Conversely, a low sense of self-regulatory efficacy increases vulnerability to social pressures for transgressive conduct (Bandura, 1991).

For the management to successfully deal with abuses, both large and small, one important task therefore is to ensure a caring and democratic climate favorable to pro-social conducts. In Freudian terms, the strength of one's ego (that regulates the primitive id impulses) is stronger in such environments than in ones that are selfish and authoritarian. People do seek social approval of their conducts, and a healthy network of interpersonal relations will reduce the possibility that one runs wild in the Internet to seek some imaginary substitute. But a caring climate is not enough, since in de-inhabitation and de-individuation one can never know if the id can be regulated at all. Using Gardner's terms, a fundamental change of the unschooled beliefs requires "Christopherian encounters," in which one must confront his or her own misconceptions. Or according to Argyris, one must practice the double-loop learning in which one's value systems must be surfaced and challenged. Or as Karl Weick suggests, one can only know what one thinks until he or she sees what he or she says. The earlier example of the two IT professionals demonstrates this practice: they only discover their misconceptions about both the Internet and women after they see what they have done.

In corporate life, however, the practice of self-monitoring and self-reflection may be discouraged. This is not because ethics is not important, but because ethics is not built into the way in which the work and the organization are structured. The division of work, the focus on efficiency, and the demand for immediate return have created "invisible individuals" who are neither knowledgeable of, nor sensitive to their respective ethical responsibility. Weick (1979) correctly points out that, in organizations, people act "thinkingly" by "sensemaking." But acting thinkingly can be unschooled, i.e., based on stereotype misconceptions, unless one is constantly engaged in retrospective reflection. Life in modern business is likely to be so hectic that it does not permit elaborate consideration. The invention and adoption of information technology so far has only worsened this trend. Furthermore, even if people become aware of ethical conflicts, they may choose explaining away the noise rather than conducting their own "Christopherian encounters." Indeed, admitting one's own deficiency may be discouraged by cultural factors, which, for example, may value seniority or face saving more than self-discovery. Thus, unless motivated and given adequate resources, knowing by acting may only reinforce what we already know, leading to "skilled unawareness and incompetence" (Argyris, 1990).

As a consequence, to safeguard workplaces from Internet abuses, both large and small, requires us to rethink the entire design of work and organizations. The Internet is such a technology that it is easily integrated into every part of our work and can connect us to the outside world. Thus, metaphorically speaking, it has the potential of connecting all of our minds: the id, ego, and superego. The Internet is therefore a sword of double edges and can both enhance and endanger our work. Yet, the industrial model and the materialism worldview still dominate our thinking when we apply this technology to the design of work and organizations. We are told the 6-D vision of info-eccentrics (Brown & Duguid, 2000): if human society consisted of a network of mechanistic minds, the world would be de-massified, decentralized, de-nationalized, de-specialized, dis-intermediated, and disaggregated. We pay attention to only the revenue growth of electronic commerce, the saving from reengineering, and the profitability of a certain dot-com. When we address work and ethics, we ignore the complexity of the mind and treat humans as if they are all utilitarian creatures. But as revealed by the evidence of Internet abuse in Taiwan, this industrial and material approach is far from adequate to guide the application of the Internet. IT professionals must now conduct their own "Christopherian encounter" to discover a way to design our work so that the pursuit of profit and the practice of sound ethics can both be attained.

Finally, the "Christopherian encounters" are needed not only for work but also for the entire society as well. Our minds, as demonstrated by both classical Freudians and modern developmental psychologists, are malleable to cultural practices and especially so when the age is young. In the meantime, the current trend of Internet adoption indicates that the Internet will be integrated into every part of our life. Certainly this may change our society fundamentally, but how? While the info-eccentrics have paid little attention to this issue, we should be aware of the grave consequence if we make some irrecoverable mistakes in making the adoption decisions for families, workplaces, and various cultures. Schooling the mind, especially at the early life of people, is more important than ever. It is not only scary, but also potentially destructive to human future if the Internet is occupied by a lot of unschooled minds that are filled with unpleasant past memories and misconceived theories of the world.

CONCLUSION

In this chapter I approach the issues pertaining to Internet abuse from a psychoanalytic perspective. To effectively confront Internet abuse, I argue, requires the knowledge of the unconscious mind. Although modern knowledge of the unconsciousness is still limited, I believe that this orientation is beneficiary to the construction of IT-laden social systems. The Internet can be a virtual world for the ids to endanger one another, or it can be a place for self-discovery that eradicates stereotype misconceptions. The outcome depends on how we view human nature and how we design work around the Internet.

This orientation also calls for new perspectives to managing human resources in modern tech-ridden companies. First, while it is important for employees to be efficient in Internet-related skills, their education must go beyond simple skill training to include courses on social responsibilities and individual psychological well-being. Also, the design of work must not ignore the importance of social interactions in the physical world. Today's design of information systems has mainly neglected the issue of social presence, which can be enhanced through office layout and interface design. As Brown and Duguid (2000) point out in their work, "virtual work" may not succeed, or may even be dysfunctional, unless socialization is an integral part of the work design. Finally, considering that social sanctions are especially important in curtailing one's primitive impulses in committing Internet abuse, companies must invest in creating and sustaining a healthy mutual-caring culture. In doing so, our goal must be to broaden our perspective to address both individuating and social issues, and to regulate both conscious and unconscious conducts. We may then be ready to confront both "abuse in the small" and "abuse in the large" effectively.

REFERENCES

Argyris, C. (1990). *Overcoming Organizational Defenses.* Needham Heights, MA: Allyn and Bacon.

Bandura, A. (1991). Social cognitive theory of moral thought and action. In Kuritines, W.M. & Gewirtz, J.L. (Eds.), *Handbook of Moral Behavior and Development, Volume 1: Theory.* Lawrence Erlbaum Associates.

Banerjee, D, Cronan, T.P., & Jones, T.W. (1998). Modeling IT ethics: A study in situation ethics. *MIS Quarterly,* 22(1), 31-60.

Brown, J.S. & Duguid, P. (2000). *The Social Life of Information.* Boston, MA: Harvard Business School Press.

Gardner, H. (1991). *The Unschooled Mind: How Children Think and How Schools Should Teach.* New York: Basic Books.

Harrington, S. (1996). The effect of codes of ethics and personal denial of responsibility on computer abuse judgments and intention. *MIS Quarterly,* 19(3), 257-278.

III. (2001). Institute of Information Industry, Taiwan. Available online at: http://www.find.org.tw.

Kleiner, A. & Roth, G. (1997). How to make your experience your company's best teacher. *Harvard Business Review*, (September/October), 172-177.

Kraut, R., Patterson, M., Lundmark, V., Kiesler, S., Mukopadhyay, & Scherlis, W. (1998). Internet paradox. *American Psychologist*, 53(9), 1017-1031.

Mitchell, S.A. & Black, M.J. (1995). *Freud and Beyond.* New York: Basic Books.

Weick, K.E. (1979). *The Social Psychology of Organizing.* Reading, MA: Addison-Wesley.

Chapter XII

Internet Abuse and Addiction in the Workplace: Issues and Concerns for Employers

Mark Griffiths
Nottingham Trent University, UK

ABSTRACT

The Internet as a communication medium has become an increasing part of many people's day-to-day working lives. As with the introduction of other mass communication technologies, issues surrounding use, abuse, and addiction have surfaced. For instance, according to a recent report carried out by the company SurfControl (Snoddy, 2000), office workers who while away one hour a day at work on various non-work activities (e.g., trading shares, booking holidays, shopping online, etc.) could be costing businesses as much as $35 million a year. The survey found that

59% of office Internet use was not work related and that those who traded in shares, played sports, shopped, and booked holidays cost companies the most. It is clear from research such as this that Internet abuse is a serious cause for concern—particularly to employers. This chapter has a number of objectives. It will first introduce readers to the concept of Internet addiction, before going on to look at the wider issue of Internet abuse in the workplace. Generic types of Internet abuse will be described, in addition to further examination of the reasons why Internet abuse occurs. The chapter ends with an overview of three very specific types of Internet abuse (i.e., online pornography, sexually related Internet crime, and online gambling), that will be of concern to employers, before concluding with some guidelines and recommendations for employers and human resources departments.

INTERNET ADDICTION: A BRIEF OVERVIEW

There have been a growing number of reports in the popular press about excessive use of the Internet under the guise of "Internet addiction," "Internet Addiction Disorder" (IAD), and "Internet Addiction Syndrome" (IAS) (Griffiths, 2000a). For many people, the concept of Internet addiction seems far-fetched, particularly if their concepts and definitions of addiction involve the taking of drugs. Despite the predominance of drug-based definitions of addiction, there is now a growing movement which views a number of behaviors as potentially addictive including those which do not involve the ingestion of a psychoactive drug (e.g., gambling, computer game playing, exercise, sex, and now the Internet) (Griffiths, 1996a).

Research has suggested that social pathologies are beginning to surface in cyberspace. These have been termed "technological addictions" (Griffiths, 1995, 1996b) and have been operationally defined as non-chemical (behavioral) addictions which involve excessive human-machine interaction. They can thus be viewed as a subset of behavioral addictions (Marks, 1990) and feature core components of addiction (Brown, 1993; Griffiths, 1996a), i.e., salience, mood modification, tolerance, withdrawal, conflict, and relapse. Young (1999) claims Internet addiction is a broad term that covers a wide variety of behaviors and impulse control problems. This is categorized by five specific subtypes:

1) *Cybersexual addiction:* compulsive use of adult websites for cybersex and cyberporn.
2) *Cyber-relationship addiction:* over-involvement in online relationships.
3) *Net compulsions:* obsessive online gambling, shopping, or day-trading.
4) *Information overload:* compulsive Web surfing or database searches.
5) *Computer addiction:* obsessive computer game playing (e.g., *Doom, Myst, Solitaire,* etc.).

In reply to Young, Griffiths (1999a, 2000a) has argued that many of these excessive users are not "Internet addicts," but just use the Internet excessively as a medium to fuel other addictions. Put very simply, a gambling addict or a computer game addict who engages in their chosen behavior online is not addicted to the Internet. The Internet is just the place where they engage in the behavior. However, in contrast to this, there are case study reports of individuals who appear to be addicted to the Internet itself (e.g., Young, 1996; Griffiths, 1996b, 2000b). These are usually people who use Internet chat rooms or play fantasy role-playing games — activities that they would not engage in except on the Internet itself. These individuals to some extent are engaged in text-based virtual realities and take on other social personas and social identities as a way of making themselves feel good about themselves.

In these cases, the Internet may provide an alternative reality to the user and allow them feelings of immersion and anonymity that may lead to an altered state of consciousness. This in itself may be highly psychologically and/or physiologically rewarding. Furthermore, as with other addictions, the activity can totally take over their life and cause many health-related problems, including both traditional withdrawal-type symptoms (e.g., moodiness, irritability, nausea, stomach cramps, etc.) and anxiety disorders, depression, and insomnia. It would appear for those with an Internet addiction disorder, the health consequences can be just as damaging as other, more traditional addictions. The good news is that the number of genuine sufferers appears to be small. However, the number will almost certainly increase over time as more and more people go online. Because of the small numbers of genuine known cases of Internet addiction, this author is unaware of very few (if any) organizations that have any practices specifically addressing this issue in the workplace (e.g., monitoring Internet addiction in the workplace, Internet addiction work policies, etc.).

There are many factors that make Internet addiction in the workplace seductive. It is clear from research in the area of computer-mediated communication that virtual environments have the potential to provide short-term comfort, excitement, and/or distraction (Griffiths, 2000a). These reasons alone provide compelling reasons why employees may engage in non-work-related Internet use. There are also other reasons, including opportunity, access, affordability, anonymity, convenience, escape, and dis-inhibition, which are outlined in more detail in the next section on Internet abuse.

Case studies of excessive Internet users may also provide better evidence of whether Internet addiction exists by the fact that the data collected are much more detailed. Even if just one case study can be located, it indicates that Internet addiction actually does exist — even if it is unrepresentative. There appear to be many people who use the Internet excessively, but are not addicted as measured by bona fide addiction criteria. Most people researching in the field have failed to use stringent criteria for measuring addiction that has perpetuated the skepticism shown among many academics. The main problems with much of the research to date is that:

- the sampling methods used have been questionable (e.g., an overreliance on self-selected samples),
- the measures used have no measure of severity,
- the measures have no temporal dimension,
- the measures have a tendency to overestimate the prevalence of problems,
- the measures used take no account of the context of Internet use,
- there is no survey work to date that conclusively demonstrates that Internet addiction exists.

Case study accounts (Griffiths, 2000b) have shown that the Internet can be used to counteract other deficiencies in the person's life (e.g., relationships, lack of friends, physical appearance, disability, coping, etc.). Most excessive Internet users spend vast amounts of time online for social contact (mostly for chat room services). As these cases show, text-based relationship can obviously be rewarding for some people and is an area for future research both in, and outside of, the workplace. As can be seen, Internet addiction appears to be a bona fide problem to a small minority of people, but evidence suggests the problem is so small that few employers take it seriously. It may be that Internet abuse (rather than Internet addiction) is the issue that employers should be more concerned about. This is therefore covered in more detail in the following sections.

TYPES OF WORKPLACE INTERNET ABUSE

It is clear that the issue of Internet abuse and Internet addiction are related, but they are not the same thing. Furthermore, the long-term effects of Internet abuse may have more far-reaching effects for the company that the Internet abuser works for than the individual themselves. Abuse also suggests that there may not necessarily be any negative effects for the user other than a decrease in work productivity.

As seen in the previous section, Young (1999) claims Internet addiction is a broad term that covers a wide variety of behaviors and impulse control problems categorized by five specific subtypes (i.e., cybersexual addiction, cyber-relationship addiction, net compulsions, information overload, and computer addiction). These can be adapted and refined to produce a typology of Internet abuse within the workplace. These are cybersexual Internet abuse, online friendship/relationship abuse, Internet activity abuse, online information abuse, criminal Internet abuse, and miscellaneous Internet abuse. These are examined in more detail below.

Cybersexual Internet abuse involves the abuse of adult websites for cybersex and cyberporn during work hours. Such online sexual services include the conventional (e.g., Internet versions of widely available pornographic magazines like *Playboy*), the not so conventional (Internet versions of very hardcore pornographic magazines), and what can only be described as the bizarre (discussion groups such as alt.sex.bondage.golden showers.sheep). There are also pornographic picture libraries (commercial and free-access), videos and video clips, live strip shows, live sex shows, and voyeuristic Web-Cam sites (Griffiths, 2000c, 2001).

Online friendship/relationship abuse involves the conducting of an online friendship and/or relationship during work hours. Such a category could also include the use of e-mailing friends and/or engaging in discussion groups, as well as maintenance of online emotional relationships. Such people may also abuse the Internet by using it to explore gender and identity roles by swapping gender or creating other personas and forming online relationships or engaging in cybersex (see above) (Griffiths, 2000c, 2001).

Internet activity abuse involves the use of the Internet during work hours in which other non-work-related activities are done (e.g., online gambling, online shopping, online travel booking, online computer gaming, online day-trading, etc.). This may be one of the most common forms of Internet abuse in the workplace.

Online information abuse involves the abuse of Internet search engines and databases. Typically, this involves individuals who search for work-related information on databases, etc., but who end up wasting hours of time with little relevant information gathered. This may be deliberate work-avoidance but may also be accidental and/or non-intentional. It may also involve people who seek out general educational information, information for self-help/diagnosis (including online therapy), and/or scientific research for non-work purposes.

Criminal Internet abuse involves the seeking out individuals who then become victims of sexually related Internet crime (e.g., online sexual harassment, cyberstalking, pedophilic "grooming" of children). The fact that these types of abuse involve criminal acts may have severe implications for employers.

Miscellaneous Internet abuse involves any activity not found in the above categories, such as the digital manipulation of images on the Internet for entertainment and/or masturbatory purposes (e.g., creating celebrity fake photographs where heads of famous people are superimposed onto someone else's naked body) (Griffiths, 2000c, 2001).

WHY DOES INTERNET ABUSE OCCUR?

There are many factors which makes Internet abuse in the workplace seductive. It is clear from research in the area of computer-mediated communication that virtual environments have the potential to provide short-term comfort, excitement, and/or distraction (Griffiths, 2000). These reasons alone provide compelling reasons why employees may engage in non-work-related Internet use. There are also other reasons (opportunity, access, affordability, anonymity, convenience, escape, dis-inhibition, social acceptance, and longer working hours) which are briefly examined below.

Opportunity and access — Obvious pre-cursors to potential Internet abuse includes both opportunity and access to the Internet. Clearly, the Internet is now commonplace and widespread, and is almost integral to most workplace environments. Given that prevalence of undesirable behaviors is strongly correlated with increased access to the activity, it is not surprising that the development of Internet abuse appears to be increasing across the population. Research into other socially acceptable but potentially problematic behaviors (drinking alcohol, gambling, etc.) has demonstrated that increased accessibility leads to increased uptake (i.e., regular use) and that this eventually leads to an increase in problems — although the increase may not be proportional.

Affordability — Given the wide accessibility of the Internet, it is now becoming cheaper and cheaper to use the online services on offer. Furthermore, for almost all employees, Internet access is totally free of charge and the only costs will be time and the financial costs of some particular activities (e.g., online sexual services, online gambling, etc.).

Anonymity — The anonymity of the Internet allows users to privately engage in their behaviors of choice in the belief that the fear of being caught by their employer is minimal. This anonymity may also provide the user with a greater sense of perceived control over the content, tone, and nature of their online experiences. The anonymity of the Internet often facilitates more honest and open communication with other users and can be an important factor in the development of online relationships that may begin in the workplace. Anonymity may also increase feelings of comfort since there is a decreased ability to look for, and thus detect, signs of insincerity, disapproval, or judgment in facial expression, as would be typical in face-to-face interactions.

Convenience — Interactive online applications such as e-mail, chat rooms, newsgroups, or role-playing games provide convenient mediums to meet others without having to leave one's work desk. Online abuse will usually occur in the familiar and comfortable environment of home or workplace, thus reducing the feeling of risk and allowing even more adventurous behaviors.

Escape — For some, the primary reinforcement of particular kinds of Internet abuse (e.g., to engage in an online affair and/or cybersex) is the sexual gratification they experience online. In the case of behaviors like cybersex and online gambling, the experiences online may be reinforced through a subjectively and/or objectively experienced "high." The pursuit of mood-modificating experiences is characteristic of addictions. The mood-modificating experience has the potential to provide an emotional or mental escape and further serves to reinforce the behavior. Abusive and/or excessive involvement in this escapist activity may lead to problems (e.g., online addictions). Online behavior can provide a potent escape from the stresses and strains of real life. These activities fall on what Cooper, Putnam, Planchon, and Boies (1999) describe as a continuum from life enhancing to pathological and addictive.

Dis-inhibition — Dis-inhibition is clearly one of the Internet's key appeals as there is little doubt that the Internet makes people less inhibited (Joinson, 1998). Online users appear to open up more quickly online and reveal themselves emotionally much faster than in the offline world. What might take months or years in an offline relationship may only takes days or weeks online. As some have pointed out (e.g., Cooper & Sportolari, 1997), the perception of trust, intimacy, and acceptance has the potential to encourage online users to use these relationships as a primary source of companionship and comfort.

Social acceptability — The social acceptability of online interaction is another factor to consider in this context. What is really interesting is how the perception of online activity has changed over the last 10 years (e.g., the "nerdish" image of the Internet is almost obsolete). It may also be a sign of increased acceptance as young children are exposed to technology earlier and so become used to socializing using computers as tools. For instance, laying the foundations for an online relationship in this way has become far more socially acceptable and will continue to be so. Most of these people are not societal misfits as is often claimed — they are simply using the technology as another tool in their social armory.

Longer working hours — All over the world, people are working longer hours and it is perhaps unsurprising that many of life's activities can be performed from the workplace Internet. Take, for example, the case of a single individual looking for a relationship. For these people, the Internet at work may be ideal. Dating via the desktop may be a sensible option for workaholic professionals. It is effectively a whole new electronic "singles bar" which, because of its text-based nature, breaks down physical prejudices. For others, Internet interaction takes away the social isolation that we can all sometimes feel. There are no boundaries of geography, class, or nationality. It opens up a whole new sphere of relationship-forming.

INTERNET ABUSE: SPECIFIC ACTIVITIES THAT EMPLOYERS SHOULD BE AWARE OF

This section briefly examines three areas (online pornography use, sexually related Internet crime, online gambling) that employers should perhaps be aware of with regards to Internet abuse by employees.

Online Pornography Use by Employees

The pornography industry was one of the first industries to take advantage of the Internet medium. It is estimated that the online pornography industry is worth $1 billion. In addition, the research company *Datamonitor* reported that sex accounts for 69% of spending on the Internet (Griffiths, 2000c). Academic researchers also claim that "sex" is the most searched for topic on the Internet (e.g., Cooper, Scherer, Boies, & Gordon, 1999; Griffiths, 2001), and as many as one-third of all Internet users visit some type of sexual site. It is also claimed that much of this activity takes place within workplace settings and is therefore an issue of major concern to employers.

All the problems that e-business and e-commerce ventures face today were first experienced by the pornography industry, which continually pushed the envelope of streaming technology because of the potential huge profits to be made. Two particular developments in current use (pay-per-click banner advertisements and real-time credit card processing) were both developed by technical expertise from within the pornographic industry. These developments have had significant impacts on the accessibility afforded to Internet users. Furthermore, theoretical 24-hour constant access has the potential to stimulate Internet abuse, which may in some circumstances lead to addictive and/or compulsive activity. Again, these factors are just as salient to those in the workplace setting as those with home Internet access.

One of the main reasons that the pornography industry has such a vested interest in this area is that in the offline world, the buying of most products is hassle-free and anonymous. However, buying pornography in the offline world may be embarrassing or stressful to the consumer, particularly if they have to go to venues deemed to be "unsavory." If pornography consumers are given the chance to circumvent this process, they invariably will. Furthermore, in the workplace setting, individuals may also be able to hide this part of their lives from their partner and/or family at home.

Sexually Related Internet Crime by Employees

The actual extent of sexually related Internet crime remains a somewhat elusive figure. However, most commentators assert that it is on the increase. The reality is that advancements in computer technology generally, and the increased availability of the Internet in particular, have provided for new innovations in, and an expansion of, the field of criminality (and more specifically in the area of sexually related Internet crime) (Griffiths, Rogers, & Sparrow, 1998).

In the broadest possible sense, sexually related Internet crime can be divided into two categories: (i) display, downloading, and/or the distribution of illegal sexually related material; and (ii) the use of the Internet to sexually procure and/or intimidate an individual in some way (e.g., online sexual harassment, cyberstalking, pedophilic grooming). Both of these are possible within the workplace, although it is likely that downloading of pornography is the most common practice within workplace settings. The police crackdown on Internet pornography has been argued by some to be futile as it could drive it underground. However, employers can introduce their own forms of crack-down in the workplace through the use of sanctions (such as wage fines or deductions, or dismissal from the job in the case of persistent offenders).

One area that has been given little consideration is that of online harassment (which is not uncommon in workplace settings). Online harassment is certainly not a new phenomenon, as there have been reported cases throughout the 1990s. For instance, in the UK, Maxine Morse gave up her £60,000-a-year job when male colleagues at the company she worked at bombarded her e-mail address with images of bestiality and naked men taken from the Internet. She was awarded £22,000 in compensation.

Electronic Harassment

In addition to Internet addiction, it is also worth highlighting the issue of online harassment and "flaming" (i.e., an abusive textual attack by another person). Such behaviors can be psychologically traumatic for the victim (Griffiths, 2001b). Words can hurt and seeing the abuse in print makes it stronger to the victim as they can read it again and again. If the post is on a list or a newsgroup, there's the added effect of knowing that lots of other people can see it, and that it's permanent. For the victims of online harassment and bullying, the health-related consequences appear to be similar to those having an Internet addiction, i.e., anxiety-related disorders, depression, and insomnia. The psychological and health effects will almost certainly impact on an employee's productivity as a result.

Online harassment and flaming can also be a pre-cursor to more serious Internet-related offences (e.g., online sexual harassment and cyberstalking). Cyberstalking is also an emerging issue that employers should be aware of. Very recently the first prosecution case of cyberstalking or harassment by computer occurred in Los Angeles when Gary Dellapenta, a 50-year-old security guard, was arrested for his online activities. It all began when Dellapenta was rebuffed by his 28-year-old victim, Randi Barber. As a result of this rejection, Dellapenta became obsessed with Barber and placed adverts on the Internet under the names "playfulkitty4U" and "kinkygal," claiming she was "into rape fantasy and gang-bang fantasy." As a result of these postings, she started to receive obscene phone calls and visits by men to her house making strange and lewd suggestions. Although such a phenomenon is by definition a global one, it was the Californian legal system that took the lead in an effort to combat it. Many other cases of cyberstalking and/or persistent and unwanted e-mail messages have also been reported, some of which have originated in the workplace.

Online Gambling by Employees

Gambling in the workplace is a little researched area despite the potential far-reaching consequences. Part of the problem stems from the fact that employers are reluctant to acknowledge gambling as a workplace issue and the possible implications that may arise from it. This section briefly examines the major issues surrounding Internet gambling in the workplace.

Internet gambling is one of the newer opportunities for gambling in the workplace. There are now a huge number of websites offering opportunities for gambling on the Internet by using a credit card. At present there are few legal restrictions to stop this form of gambling taking place. An increasing number of organizations have unlimited Internet access for all, which allows such activity to take place without arousing suspicion. Internet gambling is a somewhat solitary activity that can happen without the knowledge of both management and the employee's co-workers. Furthermore, problem Internet gambling has few observable signs and symptoms that are commonly associated with other addictions (e.g., alcoholism, drug addiction, etc.). This makes identification of problem gamblers hard for employers. However, there are a number of behaviors and "warning signs" that might be indicative of a gambling problem. Many of these involve the exploitation of time and finances.

Problem Internet gambling can clearly be a hidden activity, and the growing availability of Internet gambling is making it easier to gamble from the workplace. Thankfully, it would appear that for most people, Internet gambling is not a serious problem, although even for social Internet gamblers who gamble during work hours, there are issues about time wasting and impact on work productivity. For those whose gambling starts to become more of a problem, it can affect both the organization and other work colleagues. Managers clearly need to have their awareness of this issue raised, and once this has happened, they need to raise awareness of the issue among the work force. Employers should seek to introduce a "gambling policy" at work that includes Internet gambling. This should include a checklist that employees can assess themselves, but also include a list of behaviors and warning signs.

Finally, in this section, it might perhaps be argued that in some cases, abuse of the Internet may actually make the employee feel happier about themselves. If this is the case, it could perhaps be speculated that these individuals would actually increase (rather than decrease) productivity in the workplace. Unfortunately, there is no empirical evidence to confirm or refute such a speculation. However, it is unlikely many employers would want to facilitate Internet abuse even if it was shown that productivity could be increased in this way. There are also questions about how much Internet abuse would be acceptable and at what point the gains from feeling good start to be outweighed by excessive time spent on the Internet.

GUIDELINES FOR MANAGERS AND HUMAN RESOURCES DEPARTMENTS

As has been demonstrated, being able to spot someone who is an Internet addict or an Internet abuser can be very difficult. However, there are some practical steps that can be taken to help minimize the potential problem.

- *Take the issue of Internet abuse/addiction seriously.* Internet abuse and addiction in all their varieties are only just being considered as potentially serious occupational issues. Managers, in conjunction with Personnel Departments, need to ensure that they are aware of the issues involved and the potential risks it can bring to both their employees and the whole organization. They also need to be aware that for employees who deal with finances, the consequences of some forms of Internet abuse/addiction (e.g., Internet gambling) can be very great for the company.
- *Raise awareness of Internet abuse/addiction issues at work.* This can be done through e-mail circulation, leaflets, and posters on general notice boards. Some countries will have national and/or local agencies (e.g., technology councils, health and safety organizations, etc.) that can supply useful educational literature (including posters). Telephone numbers for these organizations can usually be found in most telephone directories.

- *Ask employees to be vigilant.* Internet abuse/addiction at work can have serious repercussions not only for the individual but also for those employees who befriend Internet abusers and addicts, and the organization itself. Fellow staff need to know the basic signs and symptoms of Internet abuse and addiction. Employee behaviors, such as continual use of the Internet for non-work purposes, might be indicative of an Internet abuse problem.
- *Give employees access to diagnostic checklists.* Make sure that any literature or poster within the workplace includes a self-diagnostic checklist so that employees can check themselves to see if they might have (or be developing) an Internet problem.
- *Monitor Internet use of your staff who you suspect may have problems.* Those staff with an Internet-related problem are likely to spend great amounts of time engaged in non-work activities on the Internet. Should an employer suspect such a person, they should get the company's IT specialists to look at their Internet surfing history as the computer's hard disk will have information about everything they have ever accessed.
- *Check Internet "bookmarks" of your staff.* In some jurisdictions across the world, employers can legally access the e-mails and Internet content of their employees. One of the most simple checks is to simply look at an employees list of "bookmarked" websites. If they are spending a lot of employment time engaged in non-work activities, many bookmarks will be completely non-work related (e.g., online dating agencies, gambling sites).
- *Develop an "Internet Abuse at Work" policy.* Many organizations have policies for behaviors such as smoking or drinking alcohol. Employers should develop their own Internet abuse policies by liaison between Personnel Services and local technology councils and/or health and safety executives.
- *Give support to identified problem users.* Most large organizations have counseling services and other forms of support for employees who find themselves in difficulties. In some (but not all) situations, problems associated with Internet use need to be treated sympathetically (and like other, more bona fide addictions such as alcoholism). Employee support services must also be educated about the potential problems of Internet abuse and addiction in the workplace.

CONCLUDING REMARKS

In this chapter, major issues that surround Internet abuse/addiction issues in the workplace have been highlighted. Internet abuse/addiction can clearly be a hidden activity, and the growing availability of Internet facilities in the workplace is making it easier for abuse to occur in lots of different forms. Thankfully, it would appear that for most people, Internet abuse is not a serious individual problem, although for large companies, small levels of Internet abuse multiplied across the workforce raises serious issues about work productivity. For those whose Internet abuse starts to become more of a problem, it can affect many levels including the individual, their work colleagues, and the organization itself.

Managers clearly need to have their awareness of this issue raised, and once this has happened, they need to raise awareness of the issue among the work force. Knowledge of such issues can then be applied individually to organizations in the hope that they can develop an Internet abuse policy in the same way that many organizations have introduced smoking and alcohol policies. Furthermore, employers need to let employees know exactly which behaviors on the Internet are reasonable (e.g., the occasional e-mail to a friend) and those which are unacceptable (e.g., online gaming, cybersex, etc.). Internet abuse has the potential to be a social issue, a health issue, *and* an occupational issue, and needs to be taken seriously by all those employers who utilize the Internet in their day-to-day business.

REFERENCES

Brown, R.I.F. (1993). Some contributions of the study of gambling to the study of other addictions. In Eadington, W.R. & Cornelius, J.A. (Eds.), *Gambling Behavior and Problem Gambling* (pp. 241-272). Reno, NV: University of Nevada Press.

Cooper, A. & Sportolari, L. (1997). Romance in Cyberspace: Understanding online attraction. *Journal of Sex Education and Therapy*, 22, 7-14.

Cooper, A., Putnam, D.E., Planchon, L.A., & Boies, S.C. (1999). Online sexual compulsivity: Getting tangled in the Net. *Sexual Addiction & Compulsivity: The Journal of Treatment and Prevention*, 6, 79-104.

Cooper, A., Scherer, C., Boies, S.C., & Gordon, B. (1999). Sexuality on the Internet: From sexual exploration to pathological expression. *Professional Psychology: Research and Practice*, 30, 154-164.

Griffiths, M.D. (1995). Technological addictions. *Clinical Psychology Forum*, 76, 14-19.

Griffiths, M.D. (1996a). Behavioural addictions: An issue for everybody? *Journal of Workplace Learning,* 8(3), 19-25.

Griffiths, M.D. (1996b). Internet "addiction": An issue for clinical psychology? *Clinical Psychology Forum*, 97, 32-36.

Griffiths, M.D. (1998). Internet addiction: Does it really exist? In Gackenbach, J. (Ed.), *Psychology and the Internet: Intrapersonal, Interpersonal and Transpersonal Applications* (pp. 61-75). New York: Academic Press.

Griffiths, M.D. (1999a). Internet addiction: Internet fuels other addictions. *Student British Medical Journa*l, 7, 428-429.

Griffiths, M.D. (1999b). Cyberstalking: A cause for police concern? *Justice of the Peace,* 163, 687-689.

Griffiths, M.D. (2000a). Internet addiction: Time to be taken seriously? *Addiction Research*, 8, 413-418.

Griffiths, M.D. (2000b). Does Internet and computer "addiction" exist? Some case study evidence. *CyberPsychology and Behavior*, 3, 211-218.

Griffiths, M.D. (2000c). Excessive Internet use: Implications for sexual behavior. *CyberPsychology and Behavior*, 3, 537-552.

Griffiths, M.D. (2001). Sex on the Internet: Observations and implications for Internet sex addiction. *Journal of Sex Research*, 38, 333-342.

Griffiths, M.D., Rogers, M.E., & Sparrow, P. (1998). Crime and IT (part II): Stalking the Net. *Probation Journal*, 45, 138-141.

Joinson, A. (1998). Causes and implications of dis-inhibited behavior on the Internet. In Gackenbach, J. (Ed.), *Psychology and the Internet: Intrapersonal, Interpersonal, and Transpersonal Implications* (pp. 43-60). New York: Academic Press.

Marks, I. (1990). Non-chemical (behavioural) addictions. *British Journal of Addiction*, 85, 1389-1394.

Snoddy, J. (2000). Bill's up for office surfers. *The Guardian,* (October 18), 28.

Young, K. (1996). Psychology of computer use: XL. Addictive use of the Internet: A case that breaks the stereotype. *Psychological Reports*, 79, 899-902.

Young K. (1999). Internet addiction: Evaluation and treatment. *Student British Medical Journal,* 7, 351-352.

Chapter XIII

Impact of Personal Internet Usage on Employee's Well-Being

Pruthikrai Mahatanankoon
Illinois State University, USA

Magid Igbaria
Claremont Graduate University, USA

ABSTRACT

The Internet has become one of most technological necessity tools in today's workplace. With the broad scope of its usefulness and its ease of use, employees find the technology most beneficial to their daily work activities as well as their personal activities. Using the established behavioral theory with data collected from Internet users in the workplace, the chapter investigates the impact of personal Internet usage on employees' job satisfaction and performance. This chapter also recommends several strategies that management can implement to increase employees' well-

being — such as Workplace Internet Usage Decision Grid, and Adaptive Internet Monitoring and Filtering Policy — while enhancing their work performance through personal Internet usage in the workplace.

INTRODUCTION

Modern organizations recognize the benefits of the Internet through its ability to communicate, research, and share essential information between employees. The Internet is the intercommunication linkage between organizations and customers, thereby creating new virtual organizational arrangements. Its usability and functionality are endless, providing future analysis of market trends, as well as competitors' moves and products, and investigating other factors that may be affecting the company's competitive position.

Since the Internet has proven to be a useful tool for businesses, many companies provide employees with access to the Internet and e-mail accounts. Despite being a productive tool, however, many employees are spending time on the Internet that is not job related during work hours. The issues of employees spending work time on personal activities are not new to management. In some ways, spending time on a personal telephone conversation, taking longer break times, or chatting with colleagues in the office is similar to personal Internet surfing. However, personal Internet usage enhances and expands the scope of personal activities beyond organizational communication norms and boundaries, which may eventually lead to the extension of non-work activities during office hours.

The research in the organizational impact of personal Internet usage in the workplace has not been investigated fully. Many managers suggest that personal web usage leads to a non-productive workforce and recommend various remedies to limit or block personal Internet usage, such as installing Internet monitoring and filtering software to filter out some unwanted websites, restricting website access, or restricting hours of access. Besides limiting personal Internet usage through technological means, some organizations also publicize an Internet Usage Policy (IUP) throughout the workplace and anticipate the policy to be one of their deterrent strategies to enhance organizational productivity. Do these actions facilitate employees' performance and job satisfaction, or do they lead to unsatisfactory and unhappy workers?

While managers and researchers are beginning to understand how the Internet can be utilized for business purposes, their understanding of the consequences of employees using organizations' Internet access for personal pleasure has provided us with mixed findings and undecided practical recommendations.

WHAT IS THE IMPACT OF PERSONAL INTERNET USAGE?

Practitioners, as well as researchers, suggest that personal Internet usage leads to negative consequences and that management should limit the viewing of leisure websites. Not only does personal Internet usage impede employees' work performance, it can be damaging to the organization in terms of increased security and infrastructure costs, network overload, and other potential risks related to the civil and legal liability of organizations (Conlin, 2000; Verespej, 2000). However, some researchers advocate that personal Internet usage may in fact lead to a positive impact on employees' well-being. Since the work environment has become more flexible, open, and autonomous, the boundaries between work and life have become more fuzzy, so that some employees are also working interchangeably both at home and at work. Others argue that organizations must take actions to empower and educate employees about the balance between work and play so that employees can utilize the Internet to its full potential (Oravec, 2002). Personal Internet usage can "facilitate the transfer of learning from the play domain to work-related tasks" (Belanger & Van Slyke, 2002, p. 65), and many excessive workplace Internet users may also be satisfied and productive workers (Stanton, 2002).

To further investigate the consequences of personal Internet usage, this chapter defines personal Internet usage as "the use of the Internet and e-mail in the workplace for personal interests." When using multi-dimensional personal web usage as identified by Mahatanankoon, Anandarajan, and Igbaria (2002), personal Internet usage behaviors can be classified into three categories: (1) personal e-commerce (PEC), (2) personal information research (PIR), and (3) personal communication (PCO). Personal e-commerce includes conducting personal investment and banking activities, and personal online shopping. Personal information research includes activities such as researching products or services related to personal interests, and reading online news, such as sports, weather, etc. Finally, personal communication involves using the Internet and web-based e-mail for non-work-related interpersonal communications. See Appendix A for details.

The purposes of this chapter are to investigate what factor leads to personal Internet usage, and to examine the impact of personal Internet usage on job satisfaction and work performance. By understanding the motivational factor and the consequences of these behaviors, the chapter then advocates practical implication of managing personal Internet usage in the workplace.

RESEARCH MODEL AND HYPOTHESES

Using the belief-attitude-behavioral model established by Ajzen (1980), the Theory of Reasoned Action (TRA) implies that employees' attitude towards Internet usage and subjective norms are the major predictors of personal Internet usage in the workplace. TRA provides a useful applicability in understanding and predicting many social behaviors (Ajzen, 1988; Fishbein & Ajzen, 1975). Since the theory has been demonstrated to be beneficial in explaining employee intentions and predicting work behavior (Becker et al., 1995), it is likely that attitude and subjective norms could be the major antecedents of personal web usage activities, together with the consequences of the behaviors — job satisfaction and work performance — all of which can be examined from this research perspective.

Figure 1. Research Model

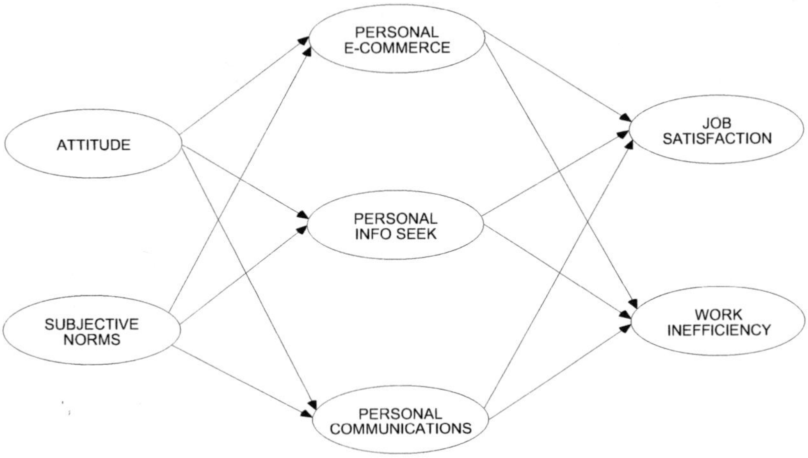

Figure 1 shows the Proposed Research Model Used in this study. According to the model, the study proposes four hypotheses:

H1: Attitude toward using the Internet (AT) is positively related to three types of personal Internet usage activities in the workplace.

Attitude toward a behavior is defined as a person's positive and negative beliefs toward performing the behavior (Ajzen, 1988). Evidence suggests that a positive attitude toward the computer influences computer usage in general (Davis, Bagozzi, & Warshaw, 1989; Klobas, 1995). If attitude toward Internet usage predicts personal web usage behaviors, then an individual will perform personal web usage activities to achieve his/her desirable outcomes. Other organizational research also suggests that attitudes toward unproductive behaviors predict employees' deviant behaviors, such as absenteeism, taking longer breaks, etc. (Bolin & Heatherly, 2001). Therefore, the attitude toward using Internet technology is defined as an individual's positive and negative feelings about using the Internet for productive or unproductive purposes, i.e., performing personal web usage activities at work. The attitude toward using Internet technology for unproductive tasks should be an important predictor of all three personal web usage activities.

H2: Subjective norms (SN) are negatively related to three types of personal Internet usage activities in the workplace.

Subjective norms are the perceptions of people important to employees regarding them in performing the behavior (Ajzen, 1991). It is how bosses and peers view employees' personal web usage at work. If the people in their companies consider personal web usage in the workplace as a negative behavior, then the employees who abide to social norms tend to avoid performing the behavior. Research in the area of computer communication media supports the thought that "social influence and local context are the key factors that determine patterns of media use" (Haythornthwaite, Wellman, & Garton, 1998, p. 211). In organizational settings, peers seem to be more influential in establishing behavioral norms in the workplace (Hollinger & Clark, 1982; Robinson & Greenberg, 1998). From the literature, it is suggested that all of the personal web usage activities are influenced more by informal peer influence rather than by formal managerial policy. The hypotheses assume that employees who are influenced by social norms are less likely to perform personal web usage activities at work.

H3: Three types of personal Internet usage activities in the workplace are positively related to job satisfaction (JS).

Personal web usage, although maybe unproductive in some cases, should have some positive effects on job satisfaction. Ang and Koh (1997) report that users who are satisfied with their informational needs are also satisfied with their jobs. Personal web usage may satisfy an individual's information need other than for just work-related purposes; therefore, it is possible that these behaviors can lead to job satisfaction. Simmers and Anandarajan (2001) find that employees who report higher levels of user satisfaction with the Internet also report that their job satisfaction has increased. The research that investigates the effects of computing in job satisfaction supports this view. Ghani et al. (1989) find that the use of personal computers has a positive effect on job satisfaction, especially when employees are working on tasks with high variety, identity, autonomy, and feedback. Baker (1995) finds that there are relationships between increasing complexity in office automation activity, and increased job motivation and job satisfaction. Zeffane (1994) suggests that the degree of job satisfaction was positively influenced by the extent of computer usage and varies by job status and functional areas. In some instances, increase in task automation may also increase employees' satisfaction (McMurtrey et al., 2002). Furthermore, there is evidence of leisure activities that lead to job satisfaction (Niehouse, 1986; Berg, 1998; Banner & LaVan, 1985; Sirgy et al., 2001). Since personal web usage is not related to employees' actual work, its behaviors can be considered as employees performing leisure activities at work. The attributes of personal web usage at work support the "spillover" hypothesis in the case where work-related web usage experience may cause employees to carry their job over into the non-work or leisure websites (Staines, 1980).

H4: Three types of personal Internet usage activities in the workplace are positively related to work inefficiency (WI).

Work inefficiency is used to identify employees' work performance. Work inefficiency refers to the time to complete work, the amount of wasted time, and the amount of re-work and extra work materials occurring from Internet usage (Anandarajan, Simmers, & Igbaria, 2000). Therefore, employees' work inefficiency should increase if employees use their Internet access for non-work-related purposes. Wen and Lin (1998) suggest that the time employees spend on personal activities reduces their productivity. Anandarajan and Simmers (2001) suggest that accessing personal-related websites at work leads to serious loss of productivity and clogged networks. Thus, personal web usage should be positively related to work inefficiency.

DATA COLLECTION AND ANALYSIS

The methodology used for this research was a web-based field survey. During the data-collection phase, e-mails were sent directly to the targeted population, asking them to complete the questionnaire. The e-mail emphasized the importance and confidentiality of the research. There were 271 respondents used in this study with an approximate response rate of 20%.[1] The respondents were 63% part-time students and 37% non-students. There were only 95 respondents who had a high school and college degrees (35%), while the rest of the respondents had a graduate or a professional degree (65%). Most of the respondents held low and middle management positions (28%) or technical positions (27%). They worked full time with an average working day close to 9.15 hours (S.D. = 1.087). There were 61% male and 39% female, with the majority of the respondents' ages ranging from 21 to 39 (69.4%).

The study used structured equation modeling (SEM) to test both the measurement model and the structural models. The measurement model consists of the relationship between the constructs and their measuring items, which need to be assessed prior to the test for significant relationships in the structural model. The structural model was examined by assessing the explanatory power of the research variables, and identifying the value and significance of the path coefficients. The path coefficient of each predictor variable (attitude and subjective norms) describes the direct effect of that variable on the mediating variable (personal Internet usage) and its consequences (job satisfaction and work inefficiency).

MEASUREMENT MODEL

The measurement model is assessed from item loadings, composite reliability, convergent and discriminant validity. All item loadings are consider acceptable; each item has a higher loading on its assigned construct than on the

Table 1. Factor Analysis and Composite Reliability

Items	ATT	SN	PEC	PIR	PCO	JS	WI
ATT1	.763						
ATT2	.861						
ATT3	.745						
SN1		.880					
SN2		.832					
SN3		.444					
PEC1			.740				
PEC2			.769				
PEC3			.712				
PEC4			.724				
PIR1				.675			
PIR2				.614			
PIR3				.781			
PIR4				.633			
PCO1					.622		
PCO2					.825		
PCO3					.462		
JS1						.901	
JS2						.842	
JS3						.821	
WI1							.702
WI2							.772
WI3							.814
WI4							.672
Mean	3.33	2.92	1.91	2.08	1.85	3.62	2.72
S.D.	.87	.83	.68	.67	.73	.88	.85
Cronbach's Alpha	.7899	.7306	.8109	.7967	.6412	.8179	.7283

other construct, based on the power level of 80% and the sample size of 200; a factor loading value of .40 or above is significant at .05 level (Hair, Anderson, Tatham, & Black, 1998). The variance extracted from the constructs ranges from .59 to .78, exceeding the .50 criterion which suggests that the constructs are distinct and unidimensional (Fornell & Larcker, 1981). In addition the composite reliabilities of each latent factor, which are analogous to estimates of coefficient alpha, ranged from .64 to .81. These values exceed the recommended values according to the guidelines (Hair et al., 1998). Table 1 shows the factor loadings and composite reliability for each latent variable. The convergent and discriminant validity can be identified through goodness-of-fit measures and c^2 (Gefen, Straub, & Boudreau, 2000). The measurement model showed a GFI of .917, an AGFI of .893, and an NFI of .879 with c^2 value significantly smaller in the proposed model, thereby supporting the convergent and discriminant validity of the measurement model.

Table 2. Results from Hypotheses Testing

Hypotheses	Latent Variables	Beta
H1	AT --> PEC	0.59***
	AT --> PIR	0.54***
	AT --> PCO	0.84***
H2	SN --> PEC	-0.37 *
	SN --> PIR	-0.20
	SN --> PCO	-0.73***
H3	PEC --> JS	0.36 *
	PIR --> JS	-0.18
	PCO --> JS	-0.16
H4	PEC --> WI	- 0.41*
	PIR --> WI	0.63**
	PCO --> WI	-0.37

*$p < 0.05$ **$p < 0.01$ ***$p < 0.001$

STRUCTURAL MODEL

Several statistics were used to assess the model's goodness of fit. The goodness of fit indices for this model included c^2/df= 1.325; Goodness of Fit Index (GFI)=.916; Root Mean Square of Approximation (RMSEA)=0.035; Comparative Fit Index (CFI)=0.966. All measures were within the acceptable levels as recommended by Bentler (1990) and Bagozzi and Yi (1988), thus indicating an acceptable structural model fit. Table 2 shows the results from the structural model.

RESULTS

The results of the multivariate test of the structural model are presented in Table 2. Hypothesis 1 was entirely supported, with attitude having significant direct effects on personal e-commerce (b=.59, p<.001), personal information research (b=.54, p<.001), and personal communications (b=.84, p< .001). Subjective norms had a significant direct effect on only personal e-commerce (b = -.37, p < .05) and personal communications (b = -.73, p < .001). In Hypotheses 3, the results showed weak support for job satisfaction, in which only personal e-commerce was supported (b=.36, p<.05). Partial support was also obtained for Hypothesis 4 where personal e-commerce ($b = -.41$, $p < .05$) and personal information research ($b = .63$, $p < .01$) had a significant direct effect on work inefficiency.

In summary, the study showed that attitude toward using the Internet was found to be the most significant factor for employees to engage in personal Internet usage activities. Subjective norms, which focused on how employees will comply with their peers and important people regarding their Internet usage behaviors, were found to have a negative impact on employees' intentions to engage in personal e-commerce and personal communications, but not on personal information seeking. Regarding the consequences of personal Internet usage, personal e-commerce enhanced job satisfaction, and interestingly, it apparently increased employees' productivity. However, personal information seeking was the only factor that decreased employees' productivity.

RECOMMENDATIONS FOR MANAGEMENT

The study provides new insight and possible strategies for managing workplace Internet infrastructure. The findings suggest that not all personal

Figure 2. Workplace Internet Usage Decision Grids

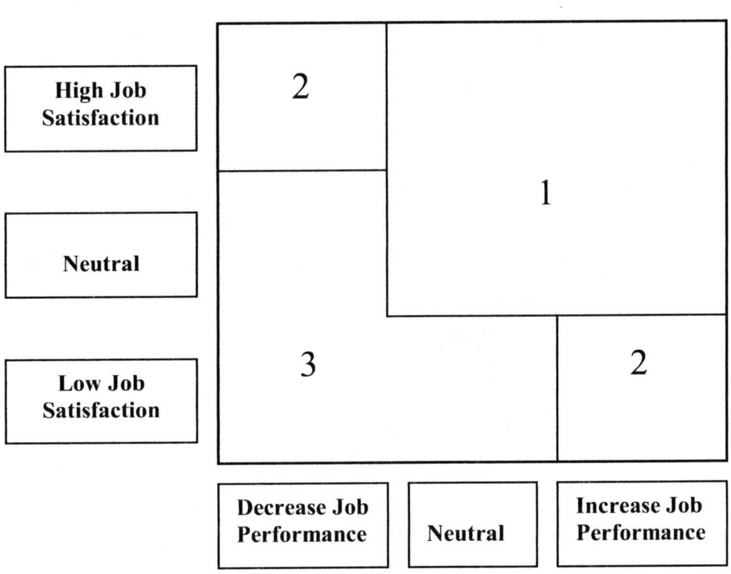

Internet usage activities lead to job satisfaction or work inefficiency. Personal e-commerce is the only activity that leads to job satisfaction; at the same time it leads to negative work inefficiency (improved productivity). Therefore, management should take provision when it comes to restricting employees' Internet usage behaviors, as any personal Internet usage activities that lead to job satisfaction also in fact increase employees' productivity. On the contrary, some personal Internet activities, such as personal information seeking, decrease productivity and do not influence job satisfaction. Employees who are extensively seeking a personal research agenda beyond their workplace duties may in fact be wasting organizational time and may not be satisfied with their jobs. Since personal information seeking has no impact on employees' job satisfaction, spending too much time on these tasks is futile for both the employees and their organizations. Personal communications have no impact on job satisfaction or work inefficiency. These are neutral activities that management has to decide if they should allow.

Figure 2 shows possible strategies that management can implement when it comes to deciding which personal Internet usage activities should be flexible and which activities should be restricted. Personal Internet activities in Area 1, such as some personal e-commerce activities, are most preferable since they lead to satisfying and productive employees. Blocking or restricting these activities can be damaging to organizational performance and employees' well-being. Whereas personal Internet activities in Area 3, such as some personal information seeking activities, should be strictly monitored or restricted, as they create lower employee work performance and have no impact on their job satisfaction. However, activities that fall within Area 2, such as some personal web-based communications, are ambiguous, and the decision regarding allowing them should be based on management judgment and trade-offs between organizational performance and employees' job satisfaction.

The results from this study also imply that management can motivate productive use of the Internet through attitudinal changes and workplace behavioral norms. Since the attitude toward Internet usage is a major predictor of personal Internet usage behaviors, management can reduce the negative effects of personal web usage behaviors through changing employee attitudes by clearly and openly communicating to them what management views as proper organizational Internet usage. Also, some personal Internet behaviors can be asserted and determined through peer behaviors, such as personal e-commerce and personal communications. However, personal information seeking, which depends on individual interests, may not be influenced by peers since most employees may not want other fellow employees to know about their personal research activities, such as job search, hobbies, and health issues.

Although management can restrict personal Internet usage through software filtering and monitoring tools, research finds that some employees are not satisfied with this policy (Urbaczewski, 2000). Therefore, we suggest that attitudinal change and enforced behavioral norms can be accomplished better through education and training. Education is necessary for employees to understand Internet usage policy, which must include how to use the Internet effectively and productively, and how to avoid abusing it. These education and training activities should include:

1) *Training regarding the general technological background of Internet technologies* — Employees' knowledge and comprehension about the nature of informational storage and the permanency of computer records are also tremendously helpful. For example, employees should understand that their e-mail messages and Internet surfing logs remain in their computer or organizations' servers for technological and administrative purposes.
2) *Training about personal consumption of information* — Employees' awareness of how to be good consumers and distributors of their own information is essentially cooperative. Employees should be responsible for their own information consumption, such as having the ability to evaluate and to screen news, information, and other advertising messages, and at the same time being responsible for their own information output by ensuring that the information that they supply is accurate, timely, and legal.
3) *Educational videos regarding Internet abuse in the workplace* — To avoid costly continuous training, organizations can create their own training videos that educate their employees on how to use the Internet and e-mail in the workplace, along with the likely consequences.
4) *Customized start-up homepage* — Organizations can also provide their employees with their organizations' customized start-up homepage that has specific hyperlinks to useful websites, thereby reducing the time employees spend surfing and searching for work-related information; even everyday information such as newspaper, weather forecasts, etc. can be provided.

Most importantly, management should understand that some personal-related Internet and e-mail usage can enhance the quality of work life and well-

Figure 3. Adaptive Internet Monitoring and Filtering Policy

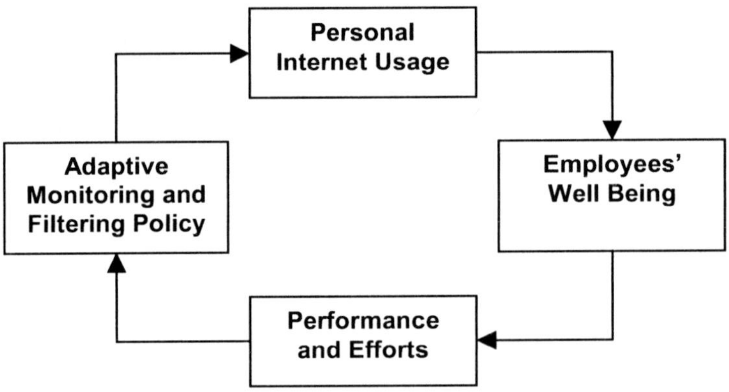

being for motivated employees; hence, organizations must take precautions against the restriction of Internet monitoring and filtering. Too much or too little control leads to Internet abuse (Anandarajan, 2002). Monitoring restrictions, based on acceptable Internet usage policy, should be based on employees' work performance and at the same time increase their well-being in the workplace. We recommend that management can maintain a healthy psychological contract of Internet usage through an *"adaptive Internet monitoring and filtering policy."* Figure 3 implies that to improve employees' well-being, organizations may allocate time for personal Internet usage, while at the same time employees should perform to their organization's expectations regarding performance and effort. This adaptive Internet monitoring and filtering policy requires a reciprocal sense of respect and fulfillment of the psychological contract between organizations and employees. It also suggests that Internet monitoring and filtering must take employees' needs and job characteristics into consideration. Other factors that can influence adaptive Internet usage policy include organizational culture, technological infrastructure, employees' roles and status, all of which generally dictate the amount of Internet usage activities.

LIMITATIONS

First, the sample used in this study consisted of a convenience sample of part-time MBA students as the majority. Other factors that may influence the nature of personal web usage activities include respondents' positions in organizations, gender, age, and education. Second, although the study has tested the measurement model, the study may not have captured all the various underlying personal Internet activities that actually exist in the workplace. As there is a huge spectrum of activities that can be performed via the Internet and e-mail, this research is based on the activities that could be considered as general personal usage norms. Management and future research need to apply our findings by examining and taking other personal Internet usage activities into consideration. And lastly, because the questionnaire asked the respondents about their non-productive behaviors of Internet usage, and because the questions were based on self-reported items, there was still a possibility that the results were somewhat biased toward positive behaviors, even if the web-based questionnaire specified clearly that it maintained anonymity and confidentiality.

CONCLUSION

The study raises new questions regarding job satisfaction, work performance, and employees' well-being in regard to their personal Internet usage. The findings show that not all personal web usage leads to work inefficiency; in some activities, it may even eventually increase job satisfaction. The study also recommends various strategies that management can implement to enhance employees' well-being, such as *Workplace Internet Usage Decision Grids*, *Adaptive Internet Monitoring and Filtering Policy*, and user education/training. These strategies will help researchers and practitioners understand possible Internet usage patterns of employees and better advise on positive Internet usage policies that fit their jobs and personal agenda.

ENDNOTES

1 The exact response rate was not known since the student mailing lists were maintained by each of the schools and there was no question in the survey to find out which school each respondent was coming from.

REFERENCES

Ajzen, I. (1988). *Attitudes, Personality, and Behavior*. Chicago, IL: The Dorsey Press.

Ajzen, I. (1991). The theory of planned behavior. *Organizational Behavior and Human Decision Processes,* 50(2), 179-211.

Anandarajan, M. (2002). Internet abuse in the workplace. *Communications of the ACM,* 45(1), 53-54.

Anandarajan, M. & Simmers, C. (2001). Factors influencing Web access behaviour in the workplace: A structural equation approach. In Anandarajan, M. (Ed.), *Internet Usage in the Workplace: A Social, Ethical and Legal Perspective* (pp. 44-66). Hershey, PA: Idea Group Publishing.

Anandarajan, M., Simmers, C., & Igbaria, M. (2000). An exploratory investigation of the antecedents and impact of Internet usage: An individual perspective. *Behavior & Information Technology,* 19(1), 69-85.

Bagozzi, P.R. & Youjae, Y. (1988). On the evaluation of structural equation models. *Journal of the Academy of Marketing Science,* 16(1), 74-94.

Belanger, F. & Van Slyke, C. (2002). Abuse or learning? *Communications of the ACM,* 45(1), 64-65.

Bentler, P. (1990). Comparative fit indexes in structural models. *Psychological Bulletin,* 107, 238-246.

Bolin, A. & Heatherly, L. (2001). Predictors of employee deviance: The relationship between bad attitudes and bad behavior. *Journal of Business and Psychology,* 15(3), 405-418.

Conlin, M. (2000). Workers, surf at your own risk. *Business Week,* (June 12), 105-106.

Davis, F.D., Bagozzi, R.P., & Warshaw, P.R. (1989). User acceptance of computer technology: A comparison of two theoretical models. *Management Science,* 35(8), 982-1003.

Fishbein, M. & Ajzen, I. (1975). *Belief, Attitude, Intentions, and Behavior: An Introduction to Theory and Research*. Boston, MA: Addison-Wesley.

Fornell, C.R. & Larcker, D.F. (1981). Structural equation models with unobservable variables and measurement error. *Journal of Marketing Research,* 18, 39-50.

Gefen, D., Straub, D.W., & Boudreau, M.-C. (2000). Structural equation modeling and regression: Guidelines for research practice. *Communications of the AIS,* 4(7), 1-77.

Hair, J.F., Anderson, R.E., Tatham, R.L., & Black, W.C. (1998). *Multivariate Data Analysis*. Upper Saddle River, NJ: Prentice Hall.

Haythornthwaite, C., Wellman, B., & Garton, L. (1998). Work and community via computer-mediated communication. In Gackenbach, J. (Ed.), *Psychology and the Internet: Intrapersonal, Interpersonal, and Transpersonal Implications* (pp. 369). San Diego, CA: Academic Press.

Hollinger, R.C. & Clark, J.P. (1982). Formal and informal social controls of employee deviance. *The Sociological Quarterly,* 23(3), 333-343.

Klobas, J.E. (1995). Beyond information quality: Fitness for purpose and electronic information use. *Journal of Information Science,* 21(2), 95-114.

Mahatanankoon, P., Igbaria, M., & Anandarajan, M. (2002). Personal Web usage in the workplace: A comparison between Thai vs. U.S. Paper presented at the *7th Asia-Pacific Decision Sciences Institution Conference,* Bangkok, Thailand.

Oravec, J.A. (2002). Constructive approach to Internet recreation in the workplace. *Communications of the ACM,* 45(1), 60-63.

Robinson, S.L. & Greenberg, J. (1998). Employees behaving badly: Dimensions, determinants and dilemmas in the study of workplace deviance. In Cooper, C.L. & Rousseau, D.M. (Eds.), *Trends in Organizational Behavior* (Vol. 5, pp. 1-30). New York: John Wiley & Sons.

Stanton, J.M. (2002). Company profile of the frequent Internet user. *Communications of the ACM,* 45(1), 55-59.

Urbaczewski, A. (2000). *An Examination of the Effects of Electronic Monitoring of Employee Internet Usage.* IN: Indiana University.

Verespej, M.A. (2000). Internet surfing. *Industry Week,* 249(3), 58-64.

Wen, H.J. & Lin, B. (1998). Internet and employee productivity. *Management Decision,* 36(6), 395-398.

APPENDIX

Scales and Items

Attitude (AT)

ATT1. Using the Internet and/or e-mail for personal-related purposes is acceptable.

ATT2. Using the Internet and/or e-mail for personal-related purposes is a wise idea.

ATT3. I like the idea of using the Internet and/or e-mail for personal-related purposes at work.

Subjective Norms (SN)

SN1. My co-workers or colleagues who influence my behavior would think that I should use the Internet and/or e-mail for work-related tasks only.

SN2. People who are important to me would think that I should use the Internet and/or e-mail for work-related tasks only.

SN3. My company would think that I should use the Internet and/or e-mail for work-related tasks only.

Job Satisfaction (JS)

JS1. Generally speaking, I am very satisfied with my job.

JS2. I frequently think of changing my job.

JS3. I am generally satisfied with the kinds of projects I do on in my job.

Work Inefficiency

Using the Internet/e-mail has resulted in...

WI1. a reduction in my time to complete work.

WI2. a reduction in the amount of time I waste.

WI3. a reduction in the amount of re-work I do.

WI4. a reduction in the amount of extraneous material I have to weed through.

Personal Web Usage (PWU)

While at your place of work, please indicate the extent of your Internet usage to perform the following activities:

Personal E-Commerce

PEC1. Conducting personal external businesses.

PEC2. Conducting personal investment and banking activities.

PEC3. Conducting personal online shopping.

PEC4. Conducting personal travel or recreational activities.

Personal Information Research

PIR1. Reading online news, including sports, weather, etc.

PIR2. Researching any products or services related to personal interests.

PIR3. Researching personal hobbies.

PIR4. Viewing entertainment products and services.

Personal Communications

PCO1. Sending e-cards, flowers, gifts, etc., to friends and family.

PCO2. Sending or forwarding e-mail to multiple mailing lists, individuals, or newsgroups.

PCO3. Using personal Web-based e-mail, such as Hotmail, Yahoo, etc.

About the Authors

Murugan Anandarajan is an Associate Professor of MIS at Drexel University, USA. His research has appeared in journals such as *Communications of the ACM, Decision Sciences, Journal of Management Information Systems, Journal of International Business Studies,* and the *Omega-International Journal of Management Science,* among others. He is a Co-Editor of the following books: *Internet Usage in the Workplace: A Social, Ethical and Legal Perspective* (2001) and *Business Intelligence in Accounting* (Springer-Verlag, 2003). He was Editor of a special section on "Internet Abuse in the Workplace" in *Communications of the ACM* (December 2001).

Claire A. Simmers received her PhD from Drexel University, Philadelphia, Pennsylvania, USA, in Strategic Management. She was recently promoted to Associate Professor in the Management Department in the Erivan K. Haub School of Business at Saint Joseph's University, Philadelphia. She teaches courses at the undergraduate, MBA, and executive level in Business Policy, International Management, Leadership, and Managerial Skills. Her research interests are in political and behavioral influences in strategic decision-making, work/life issues, and the socio-technical interface in the digital economy, focusing on the Internet. Her work has been published in *Behaviour and Information Technology*, *The Journal of Business and Economics Studies*, *Communications of the ACM*, *Journal of Information Technology, Theory and Application,* and the *Journal of Organizational Behavior.* She co-authored a book with Murugan Anandarajan, PhD, titled, *Managing Web Usage in the Workplace: A Social, Ethical and Legal Perspective,* published by Idea Group in 2002. She is a Contributing Author in Porth, S.J. (2002). *Strategic Management: A Cross-Functional Approach.* Upper Saddle River, NJ: Prentice-Hall.

* * *

Michael Aikenhead holds a degree in Computer Science and several degrees in Law. After completing studies at the University of Melbourne in Australia, Dr. Aikenhead moved to England to pursue research both into the emerging area of information technology law, and into the uses of technology for transforming legal and governmental processes. He completed his doctorate at the Centre for Law and Computing at the University of Durham (UK) and was awarded a lifetime honorary fellowship of the Centre in 2001. He has published widely on numerous aspects of technology law; has worked and lectured in Australia, Europe, and the United States; and currently works for a multi-national technology company designing IT solutions for e-government.

Paulette S. Alexander is Associate Professor and Chair of the Department of Computer Information Systems at the University of North Alabama, USA. Her PhD is in Information Systems from The University of Memphis, Fogelman College of Business and Economics. She holds a Certificate in Data Processing from the Institute for Certification of Computer Professionals. She also holds a Master's of Public Affairs from The University of Texas at Austin, Lyndon B. Johnson School of Public Affairs, and Master's of Arts and Bachelor's of Science degrees from The University of Alabama, Tuscaloosa. Dr. Alexander has been on the faculty of the University of North Alabama (USA) since 1981. Her primary research and teaching interests are in the areas of Internet privacy and information systems management. She would like to express appreciation to Shruti Jalan Gupta, who served as research assistant for this project.

Grania Connors is a lawyer who specializes in information technology and intellectual property law. She has completed a Master's of Laws (Research) on the international treatment of automated electronic contracting and holds a Master's of Technology in Computing. Ms. Connors has acted as a consultant to government, advising on the impact of technology in the legal profession. She has published several legal works relating to privacy and breach of confidence by employees, in addition to a computer maintenance handbook. She currently works as a Legal Analyst, advising an international software group on interpreting and implementing statutory law.

Patrick Devine is a graduate student pursuing his doctorate in Management Information Systems at Drexel University, USA. He received his BS and MBA from St. Joseph's University, where he received the Graduate Business Award. His research interests include: e-commerce; Internet integration in the workplace; and the social, ethical, and legal dimensions of Internet abuse. In addition to co-authoring a chapter in the book, *Managing Web Usage in the Workplace,* his research has been or will be forthcoming in proceedings including the International Conference on Information Systems, The Annual Meeting of the Academy of Management, and the British Academy of Management.

Mark Griffiths is a Professor of Gambling Studies at the Nottingham Trent University, UK. He is internationally known for his work in gambling and gaming addictions, and was the first recipient of the John Rosecrance Research Prize for "Outstanding Scholarly Contributions to the Field of Gambling Research" in 1994 and the winner of the 1998 CELEJ Prize for best paper on gambling. He has published more than 110 refereed research papers, two books, numerous book chapters, and more than 250 other articles. His current interests are concerned with technological addictions, particularly computer games and the Internet.

The late **Magid Igbaria** was a Professor of Information Science at the Claremont Graduate University (USA) and at the Faculty of Management, Graduate School of Business, Tel Aviv University. Formerly, he was a Visiting Professor of Decision Sciences at the University of Hawaii in Manoa and a Professor of MIS in the College of Business and Administration at Drexel University. He has published articles on virtual workplace, information economics, computer technology acceptance, IS personnel, management of IS, compumetrical approaches in IS, and international IS in *Communications of the ACM, Computers and Operations Research, Decision Sciences, Deci-*

sion Support Systems, Information and Management, Information Systems Research, Journal of Management Information Systems, Omega, Journal of Strategic Information Systems, MIS Quarterly, and others. His recent research interests focused on electronic commerce, the virtual workplace, telework, computer technology acceptance, information and computer economics management of IS, IS personnel, and international IS. He served on the editorial board of *Information Resources Management Journal, Journal of the Association for Information Systems, Journal of Management Information Systems, Journal of Engineering and Technology Management, Journal of End-User Computing, Information Technology and People,* and *Computer Personnel.* He also was an Associate Editor of *ACM Transactions on Information Systems, Journal of Information Technology Cases and Applications,* and *MIS Quarterly*, and served on the executive committee of Information Systems-HICSS-30 (1997), HICSS-31 (1998), and HICSS-32 (1999). He was co-author of *The Virtual Workplace* (Idea Group Publishing, 1998). Professor Igbaria died on August 3, 2002, of complications from cancer at the age of 44.

Yongbeom Kim is an Associate Professor of Information Systems at Fairleigh Dickinson University (USA). He received his BS and MS in Electrical Engineering from Seoul National University, and MPhil and PhD in Information Systems from the Stern School of Business of New York University. His research interests include software reuse, performance evaluation of interactive computer systems, and enterprise resource planning. His work has been published in the *Journal of Management Information Systems, Information Processing & Management,* and *Journal of Information Systems Education.*

Feng-Yang Kuo holds a PhD in Information Systems from the University of Arizona. He was a faculty member of Information Systems at the University of Colorado at Denver from 1985 to 1997, and is currently a Professor of Information Management in National Sun Yet-Sen University, Taiwan. Professor Kuo's research interests include information ethics, cognition and learning in organizations, and human-computer interactions. He has published articles in *Communications of ACM, MIS Quarterly, Journal of Business Ethics, Information & Management, Journal of Systems and Software*, and *Decision Support Systems*.

Younghwa Lee is a doctoral candidate at the Leeds School of Business, University of Colorado at Boulder, USA. He received BA and MBA degrees from Korea University. His research interest is in Web usability, technology acceptance, and security. His research has been published in *Communications of the ACM, Computers & Security, Information Management and Computer Security*, and in several conference proceedings, including *International Conference on Information Systems (ICIS)*, *Academy of Management (AoM),* and *Americas Conference on Information Systems (AMCIS)*.

Zoonky Lee is an Assistant Professor of Information Systems at the University of Nebraska - Lincoln, USA. His research interests include designing IT infrastructures for organizational learning, electronic business, and IT support for knowledge workers. He has published in various journals, including *Information and Management*, *Journal of Organizational Computing and Electronic Commerce*, *Electronic Markets*, *Journal of Information Technology*, *Communications of the ACM, Computer and Security,* and *Journal of Business Strategies*.

Susan K. Lippert is an Assistant Professor of Management Information Systems in the Department of Management at Drexel University, Philadelphia, Pennsylvania, USA. Her current research interests include use and management of information technology, with an emphasis on technology trust. Dr. Lippert received her PhD in MIS and an MBA in Logistics, Operations, and Materials Management from The George Washington University, Washington, DC. Dr. Lippert has published in *Communications of the ACM*, *Journal of End User Computing*, *Journal of Management Education*, *Journal of Mathematics and Science Teaching,* and *Annals of Cases on Information Technology*.

Pruthikrai Mahatanankoon is an Assistant Professor of Information Systems in the Applied Computer Science Department at Illinois State University, USA. He holds a bachelor's degree in Computer Engineering from King Mongkut's University of Technology Thonburi, Thailand, and MS degrees in Management Information Systems and Computer Science from Fairleigh Dickinson University. He earned his PhD in Management Information Systems from the Claremont Graduate University. He has published articles in *International Journal of Electronic Business, Encyclopedia of Information Sys-*

tems, DSI proceedings, and other academic book chapters. His current research interests focus on Internet technology usage and abuse in the workplace, mobile commerce, Web services, quantitative research methods, and virtual workplace and virtual organizations.

Dinesh A. Mirchandani is an Assistant Professor of Information Systems in the College of Business Administration at the University of Missouri - St. Louis, USA. He earned his PhD in Business Administration (MIS) from the University of Kentucky in 2000, an MS in Electrical Engineering from Purdue University in 1994, and a bachelor's degree in Electronics Engineering (with honors) from the University of Bombay in 1991. His research interests include information systems planning and electronic business. His papers have been published in academic journals such as *Communications of the ACM*, *Journal of Organizational Computing and Electronic Commerce*, *International Journal of Electronic Commerce*, *Information & Management*, and *International Journal of Production Economics*, among others.

Jo Ann Oravec is an Associate Professor in the College of Business and Economics at the University of Wisconsin at Whitewater (USA). She received her MBA, MA, MS, and PhD from the University of Wisconsin at Madison. She taught computer Information Systems and Public Policy at Baruch College of the City University of New York, and also taught in the School of Business and the Computer Sciences Department of UW-Madison. In the 1990s, she chaired the Privacy Council of the State of Wisconsin, the nation's first state-level council dealing with information technology and privacy concerns. She has written several books and dozens of articles on computing technology issues, and has also worked for public television and developed software along with her academic ventures.

Andrew Urbaczewski is an Assistant Professor of Management Information Systems at the University of Michigan - Dearborn, USA. He earned a PhD from Indiana University in 2000. His work has appeared in *Communications of the ACM*, *Journal of Organizational Computing and Electronic Commerce*, *Business Horizons*, and other leading journals and conferences. He is a frequently invited professor at several universities in Europe, where he has given courses on telecommunications policy and practice, electronic commerce, and IT project management.

Index

D

E

F

G

H

I

J

K

L

T

U

W